Instrumentation and process measurements

W. BOLTON

Longman Scientific & Technical

Longman Scientific & Technical,
Longman Group UK Limited,
Longman House, Burnt Mill, Harlow,
Essex CM20 2JE, England
and Associated Companies throughout the world.

First published 1991

British Library Cataloguing in Publication Data

Bolton, W. (William), *1933–*
  Instrumentation and process measurements
  1. Instrumentation systems
  I. Title
  620.0044

Produced by Longman Group (FE) Limited
Printed in Hong Kong

Contents

Preface

What the book aims to do

This book has the aims of:

- introducing readers to the basic elements of instrumentation systems;
- enabling readers to develop a basic understanding of the techniques used for the measurement of the process variables of pressure, level, density, flow and temperature;
- enabling readers to appreciate the need for maintenance of measurement systems.

The format of the book

The chapters in the book are arranged as follows:

Overview of instrument systems	Chapter 1
The constituent elements of such systems	Chapters 2, 3, 4
Measurement techniques for process variables	Chapters 5, 6, 7, 8
Maintenance of instrument systems	Chapter 9

Each chapter includes worked examples and problems, answers appearing at the end of the book.

Courses for which suitable

The book more than covers the BTEC First Certificate/ Diploma in Engineering unit Instrumentation and Process Measurements (F 28718). It is also seen as being of value to students on other BTEC courses and CGLI students, indeed any course where a basic introduction to instrumentation systems and process measurements is required.

Background knowledge assumed   A basic knowledge of science and mathematics has been assumed. The science required is basic physical science. Key elements of science are, however, developed in the appropriate chapters. The mathematics required is just the ability to handle numbers, interpret graphs and in a few instances handle simple algebraic equations.

W. Bolton

1 Basic instrument systems

There are, in engineering, essentially three different ways instrumentation is used for making measurements:

1 *Obtaining data about some event or item* This could, for instance be the marking out of an item for machining and involve measurements of lengths and angles.

2 *Inspecting an item to see if it matches the specification* In production engineering measurements are often concerned with determining whether an item being produced has the right dimensions, shape, electrical resistance, etc., that have been specified for the item. This might mean measuring its length to see if it is the right size. Such measurements are often made on samples taken from a production line and on the basis of such measurements the entire production run is passed as being to specification. This form of inspection can be made manually by someone taking a component and actually making the measurements or automatically by test equipment fitted to the production line.

3 *Making measurements to ensure that a process is kept under control* Many industrial processes are continuous. The purpose of measurements in such a situation is to ensure the proper control of the process. For example, a process may involve taking a hot liquid from a tank. The level of the liquid in the tank and the temperature have to be continually monitored. Such types of measurement are known as *process measurements*.

This book is primarily directed at measurement systems for process measurements. However Chapters 1, 2, 3, 4 and 9 are concerned with the basic elements common to all measurement systems and thus apply to all the above forms of measurement. Chapters 5, 6, 7 and 8 are concerned with the measurement of pressure, level, density, flow and temperature. These measurements are the key ones in industrial processes.

Measurement systems

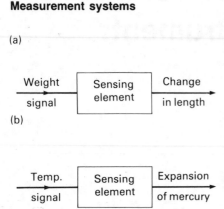

Fig. 1.1 Examples of the sensing elements with their inputs and output signals, (a) a spring balance, (b) a mercury-in-glass thermometer

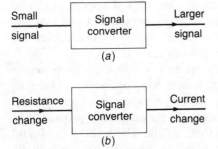

Fig. 1.2 Examples of signal converters with their input and output signals, (a) an amplifier, (b) an electrical circuit

All measurement systems consist essentially of three main parts.

1 *The sensing element* This, frequently called the *transducer*, is the first element. It is in some way 'in contact' with what is being measured and produces some signal which is related to the quantity being measured (Fig. 1.1). For example, with a spring balance used to measure weight the sensing element can be considered to be the stretching of the spring as a result of the weight on the balance. Sensing elements take information about the thing being measured and change it into some form which enables the rest of the measurement system to give a value to it. Thus the spring balance takes information about the weight and changes, into a change in length of a spring.

Change in weight $\longrightarrow$ change in length

2 *The signal converter* The output from the sensing element then passes through a second element before reaching the display. This second element can take a number of forms. In general it can be considered the *signal converter* in that the signal from the sensing element is converted into a form which is suitable for the display or control element (Fig. 1.2). An example of this might be an amplifier which takes a small signal from the sensing element and makes it big enough to activate the display.

Small signal $\longrightarrow$ bigger signal

3 *The display* The third part of the measuring system could be a display or control system. The *display* element is where the output from the measuring system is displayed. This may, for example, be a pointer moving across a scale. The display element takes the information from the signal converter and presents it in a form which enables an observer to recognise it. The *control system* is where the output from the measuring system is used to control a process. It could, for example, be used to open or close a valve to allow liquid into a container.

Measurement systems can be represented by a block diagram, with blocks to represent each of the constituent parts (Fig. 1.3). The sensing element takes a signal from the quantity being measured and changes it into a form suitable for the measurement system. The signal converter takes that signal and converts it into a suitable form for the display. In the case of a measurement instrument the display element then presents the signal in a suitable form for the observer.

We can represent the sequence of signal changes for a measurement system as:

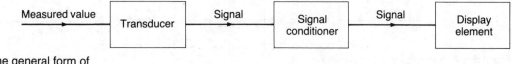

Fig. 1.3 The general form of measurement systems

signal from quantity being measured
⟶ signal suitable for the measurement system
⟶ signal suitable for display
⟶ displayed signal

Examples of measurement systems

A resistance thermometer depends for its action on a sensing element which changes resistance when the temperature changes. This resistance change might then be converted to a current change and this then used to move a pointer across the scale of a meter or perhaps a pen across a chart. Figure 1.4 shows the basic form of such a system (see Ch. 8 for more details). The signal changes for such a system are:

change in temperature
⟶ change in resistance
⟶ change in current
⟶ movement of pointer across a scale

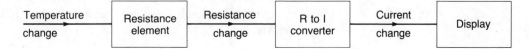

Fig. 1.4 A resistance thermometer

A thermocouple is a sensing element which gives a small voltage which is related to its temperature. This voltage is very small and thus an amplifier might be used to make the voltage signal bigger before it is used to move a pointer across the scale of a meter or perhaps a pen across a chart. Figure 1.5 shows the basic form of such a system (see Ch. 8 for more details). The signal changes for such a system are:

change in temperature
⟶ change in voltage
⟶ bigger change in voltage
⟶ movement of a pointer across a scale

Performance terms

The following are some of the terms commonly used to describe the performance of measurement systems.

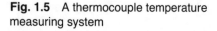

Fig. 1.5 A thermocouple temperature measuring system

Accuracy

The *accuracy* of an instrument is the extent to which the reading it gives might be wrong. A spring balance might, for example, be said to have an accuracy of ± 5 g. This means that when the spring balance is used to measure the weight of an object it can only be stated as lying within plus or minus 5 g of the balance reading. Thus a reading of 50 g means that the value of the measured quantity is somewhere between 45 and 55 g.

Accuracy is often quoted as a *percentage of the full-scale deflection (f.s.d.)* of the instrument. Thus, for example, an ammeter might have a full-scale deflection of 5 A and an accuracy quoted as ± 5%. This means that the accuracy of a reading of the ammeter between 0 and 5 A is plus or minus 5% of 5 A, i.e. plus or minus 0.25 A. Hence if the reading was, say, 2.0 A then all that can be said is that the actual current lies between (2.0 − 0.25) A and (2.0 + 0.25) A (Fig. 1.6).

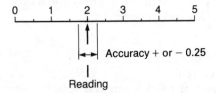

Fig. 1.6 An accuracy of ± 5% f.s.d.

It is often necessary in quoting accuracies to distinguish between whether the measurement is being made of slowly or quickly changing quantities. The term *static accuracy* is used when the quantity being measured is either not changing or changing very slowly, *dynamic accuracy* when it is changing quickly.

Error

The *error* of a measurement is the difference between the result of the measurement and the true value of the quantity being measured.

Error = measured value − true value

Thus if the value given by a thermometer is 34 °C when the true value of the temperature is 33 °C then the error is 1 °C. See later in this chapter for a discussion of the sources of error.

Repeatability

The *repeatability* of an instrument is its ability to display the same reading for repeated applications of the same value of the quantity being measured. Thus, for example, if an ammeter was being used to measure a constant current and gave for four successive readings 3.20 A, 3.15 A, 3.25 A, and 3.20 A then there can be an error with any one reading due to a lack of repeatability.

Reliability

The *reliability* of an instrument is the probability that it will operate to an agreed level of performance under the conditions specified for its use. Thus, for example, an instrument for measuring a load might have an accuracy of $\pm$ 4% now and still $\pm$ 4% six months later, while a less reliable instrument might have deteriorated and the accuracy become $\pm$ 5%.

Reproducibility

The *reproducibility* or *stability* of an instrument is its ability to display the same reading when it is used to measure a constant quantity over a period of time or when that quantity is measured on a number of occasions.

Sensitivity

The *sensitivity* of an instrument is given by:

$$\text{sensitivity} = \frac{\text{change in instrument scale reading}}{\text{change in the quantity being measured}}$$

Thus, for example, a voltmeter might have a sensitivity of 1 scale division per 0.05 V. This means that if the voltage being measured changes by 0.05 V then the reading of the instrument will change by one scale division.

Resolution

The *resolution* or *discrimination* of an instrument is the smallest change in the quantity being measured that will produce an observable change in the reading of the instrument. One factor that determines the resolution is how finely the scale of the display is divided into subdivisions. Thus if a thermometer has a scale marked in just 1°C intervals then it is not possible to estimate a temperature more accurately than about quarter or half a degree.

Range

The *range* of an instrument is the limits between which readings can be made. For example, an ammeter might have a range of 0–3 A. This means it can be used to measure currents between 0 A and 3 A. A thermometer might have a range of −10°C to 110°C and can thus be used to measure temperatures between −10°C and 110°C.

Dead space

The *dead space* of an instrument is the range of values of the quantity being measured for which it gives no reading. Also see the following definition of threshold.

Threshold

When the quantity being measured is gradually increased from zero a certain minimum level might have to be reached before the instrument responds and gives a detectable reading. This is called the *threshold*. It is just a dead space that happens to occur when the instrument is used from a zero value (Fig. 1.7). Thus, for example, a pressure gauge might not respond until

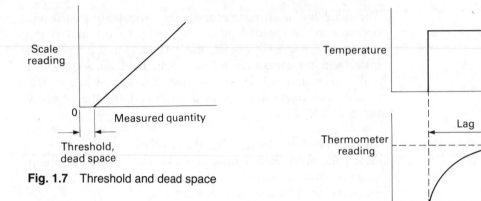

Fig. 1.7 Threshold and dead space

Fig. 1.8 Lag

the pressure has risen to about 1 kPa. This may be because of friction in the workings of the gauge.

Lag

When the quantity being measured changes a certain time might have to elapse before the measuring instrument responds to the change (Fig. 1.8). It is said to show *lag*. For example, if a mercury-in-glass thermometer is put into a hot liquid there can be quite an appreciable time lapse before the thermometer indicates the actual temperature. This is because it takes heat a significant amount of time to travel through the glass of the thermometer and increase the temperature of the mercury.

Hysteresis

Instruments can give different readings for the same value of measured quantity according to whether that value has been reached by a continuously increasing change or a continuously decreasing change. This effect is called *hysteresis* and it occurs as a result of such things as bearing friction and slack motion in gears in instruments. For example, a Bourdon pressure gauge (see Ch. 7) used to measure a pressure difference of 40 kPa can give different readings if the pressure is increased from zero to 40 kPa than if it is reduced from some higher value down to 40 kPa. Figure 1.9 shows the type of relationship that can occur for an instrument showing hysteresis.

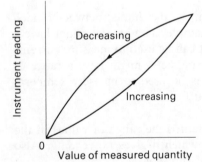

Fig. 1.9 Hysteresis, the readings being different when the measured quantity is increasing from when decreasing

Example 1

A hydrometer is specified as having a range of 600–650 kg/m^3 and an accuracy of $\pm$ 0.5 kg/m^3. Explain the significance of this data.

Answer

The hydrometer can be used to measure densities provided they are between 600 kg/m^3 and 650 kg/m^3. The results of a measurement can only be said to be lying between plus or minus 0.5 kg/m^3 of the reading given by the instrument.

Example 2

An ammeter is quoted has having a range of 0–3 A and an accuracy of ± 2% f.s.d. What is the accuracy which can be quoted for a current reading of 2 A?

Answer

The accuracy of any reading in the range 0–3 A is plus or minus 2% of 3 A, i.e. 0.06 A. Hence a current reading of 2 A has an accuracy of ± 0.06 A.

Example 3

A thermocouple is specified as having a sensitivity of 0.03 millivolt/°C. What does this mean?

Answer

When the temperature changes by 1°C the thermocouple voltage will change by 0.03 millivolt.

Example 4

A force-measuring system has a range of 0–200 N with a resolution of 0.1% f.s.d. What is the smallest change in force that can be measured?

Answer

The smallest measurable change is:

$$\frac{0.1}{100} \times 200 = 0.2 \text{ N}$$

Sources of error

The following are some of the more common sources of error that can occur with measurement systems.

Construction errors

These errors arise from such causes as tolerances on the dimensions of components and electrical components used in the manufacture of an instrument. Thus, for example, a resistor for use in a meter might be specified as being acceptable if it is within ± 1% of the required value of 100 Ω. This means that some instruments might have a resistor with a value of 99 Ω and some 101 Ω. This variation between instruments will lead to differences in readings and hence errors if all have been assumed to give the same reading.

Non-linearity errors

For many instruments a linear scale is used. Figure 1.10(*a*) shows an example of a linear scale and Fig. 1.10(*b*) a non-linear scale. A linear scale means the reading given is directly proportional to the distance or angle moved by a pointer across the scale. For an ammeter which has a linear scale, when the current is doubled the pointer sweeps through twice the angle. Figure 1.11 shows the form of a graph of instrument reading against quantity being measured when the relationship is linear. The graph is a straight line graph passing through the zero reading – zero quantity point. In many instances however, though a linear scale is used the relationship is not perfectly linear and so errors occur.

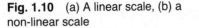

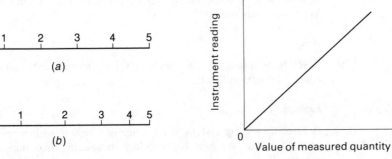

Fig. 1.11   The relationship for a linear scale

Fig. 1.10   (a) A linear scale, (b) a non-linear scale

Operating errors

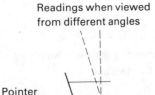

Fig. 1.12   Parallax errors

These can result from a variety of causes. *Parallax errors* are common with instruments that have pointers moving across scales, and result from the scale and pointer not being in the same plane, so the reading obtained would depend on the angle at which the pointer is viewed against the scale (Fig. 1.12). One way of ensuring that the pointer and scale is always viewed from the same angle is to have a mirror in the plane of the scale. The pointer is then always viewed in direct alignment with its image in the mirror. With a pointer position being read from a scale there are also resolution errors due to the uncertainty that exists in reading an instrument's display (see under 'Resolution' in earlier discussion).

Environmental errors

These are errors which can arise as a result of environmental effects which are not taken account of, e.g., a change in temperature affecting the value of a resistance. The specification for many instruments includes a statement of the temperature at which its readings are to the quoted accuracy. A change in temperature may result in a zero drift and/or a change in sensitivity (Fig. 1.13). Zero drift is when the zero reading of an instrument changes. Thus, for example, a meter which might have its pointer on the zero mark one day might at some other temperature, or time, indicate a small reading despite still not being used to make a measurement.

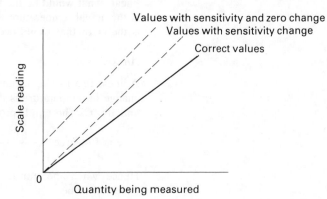

Fig. 1.13 Effect on readings of a change in temperature

Ageing errors

A consequence of instruments getting older is that some components may deteriorate and their values change; also a build-up of deposits may occur on surfaces which can affect contact resistances and insulation.

Insertion errors

These are errors which result from the insertion of the instrument into the position to measure a quantity affecting its value. For example, inserting an ammeter into a circuit to measure the current will change the value of the current due to the ammeter's own resistance.

Example 5

A pressure gauge is quoted as having a combined non-linearity and hysteresis error of ± 0.5% of any value indicated. What is the accuracy when the gauge gives a reading of 120 kPa?

Answer

The accuracy of a reading of 120 kPa is plus or minus 0.5% of that value, i.e. 6 kPa. The reading can thus be stated as 120 ± 6 kPa.

Example 6

A vapour pressure thermometer is specified as having a non-linear scale. What does this mean?

Answer

The scale marks are not equally spaced for equal changes in temperature. Thus, for example, the spacing between the scale marks for 10 and 20°C will not be the same as that between 20 and 30°C.

Example 7

A thermocouple gives no e.m.f. at 0°C and 0.645 mV at 100°C. If the relationship between e.m.f. and temperature was assumed to be linear, what would be the temperature when the e.m.f. is 0.173 mV? If the actual temperature when this e.m.f. is produced is 30°C, what is the error that would occur from assuming a linear relationship?

Answer

A linear relationship means that the e.m.f. produced for a one-degree change in temperature is the same for all temperatures between 0°C and 100°C. The e.m.f. produced per degree is

$$\text{e.m.f./degree} = \frac{0.645 - 0}{100} = 0.00645 \text{ mV}$$

Hence when the e.m.f. is 0.173 mV the temperature would be expected to be

$$\text{temperature} = \frac{0.173}{0.00645} = 26.8°C$$

The non-linearity error is thus $30 - 26.8 = 3.2°C$.

Example 8

An electrical circuit consists of a 6.0 V battery, of negligible internal resistance, with a 100 Ω resistor across its terminals. (a) What is the current through the resistor? (b) What would be the current indicated by an ammeter, of resistance 50 Ω, which was connected in series with the resistor?

Answer

(a) Using the relationship $V = IR$, then

$$\text{current } I = \frac{V}{R} = \frac{6.0}{100} = 0.06 \text{ A}$$

(b) For resistors in series the total resistance is the sum of the individual resistances. Hence the resistance across the battery when the meter is in place is $100 + 50 = 150$ Ω. Hence

$$\text{current } I = \frac{V}{R} = \frac{6.0}{150} = 0.04 \text{ A}$$

Inserting the meter has thus changed the current by 0.02 A.

Random and systematic errors

Errors can be classified as being either random or systematic errors. *Random errors* are ones which can vary in a random manner between successive readings of the same quantity. They might be due to operating errors, e.g., instruments being read at different angles and so giving variable parallax errors, or environmental errors, e.g., changes in the temperature of the surroundings affecting the calibration of the instrument. Any one reading which is subject to random errors will be inaccurate. However, to some extent random errors can be overcome by repeated readings being taken and an average calculated.

Thus, for example, suppose the constant level of water in a tank is measured a number of times and the following results obtained:

   1.010   1.015   1.010   1.005   1.010 metres

The average is obtained by adding together the results and dividing by the number of results, i.e.

$$\frac{1.010 + 1.015 + 1.010 + 1.005 + 1.010}{5}$$

Hence the average is 1.010 m. The results all lie within ± 0.005 m of this average. Hence the level would be quoted as 1.010 ± 0.005 m.

The term *precision* is generally used to describe the closeness of the agreement occurring between the results obtained for a quantity when it is measured several times under the same conditions. It is a measure of the scatter of results obtained from measurements as a result of random errors. A precise measurement system gives less scatter than a less precise system.

Systematic errors are errors which do not vary from one reading to another. They may be the result of construction or non-linearity errors and so indicated in the accuracy stated for an instrument by the manufacturer. There are also other sources of systematic errors which will not have been allowed for in the specified accuracy, e.g. a bent meter needle and insertion errors.

Example 9

What is the average result and its accuracy for a temperature when repeated measurements gave the following results:

   60.5   61.0   61.0   60.5   61.5   61.0°C

Answer

The average is the sum of the results divided by the number of results, hence

$$\text{average} = \frac{60.5 + 61.0 + 61.0 + 60.5 + 61.5 + 61.0}{6}$$

$$= 60.9\,°C$$

All the results lie within $\pm\,0.6\,°C$ of the average and so the result can be quoted as $60.9 \pm 0.6\,°C$.

Calibration

Calibration can be defined as the process of determining the relationship between the values of the quantity being measured and what the instrument indicates. Calibration of an instrument can be carried out by comparing its readings with those given by another instrument that has been adopted as a standard or by checking its readings with values specified for items, e.g., checking a voltmeter reading against the e.m.f. of a cell that is specified as giving a certain e.m.f.

The results of such a calibration might be a table giving the values of the measured quantities and the corresponding readings given by the instrument. For example, calibration of an ammeter might give the results shown in Table 1.1. The term deviation is used for the difference between the value indicated by the instrument and the true value. Thus, when the scale reading is 1.0 the actual current is $1.0 + 0.02$ A, when 2.0 it is $2.0 - 0.01$ A.

When an instrument is purchased calibration data is generally supplied. Instrument manufacturers have sets of reference instruments against which all the instruments they produce are calibrated. From time to time their reference instruments are sent to a calibration centre for checking. Their instruments in turn are calibrated against national standards.

In many companies some instruments are kept in a company standards department and used solely for calibration purposes to check the instruments used on production lines. In some cases companies keep standard length bars or blocks for checking the calibration of instruments. Thus the relationship between the calibration of an instrument in everyday use and the national standards is likely to be as shown in Fig. 1.14.

Table 1.1 Calibration data for an ammeter

Scale reading (A)	Deviation (A)
0	0
0.5	+0.01
1.0	+0.02
1.5	+0.01
2.0	−0.01

Example 10

The following data was obtained from the calibration of a thermometer. What will be the maximum error when the thermometer is used?

Reading °C	0	10	20	30	40	50
Deviation °C	0	+0.2	+1.0	+1.4	+0.8	0
Reading °C	60	70	80	90	100	
Deviation °C	−0.4	−1.5	−1.6	−0.2	0	

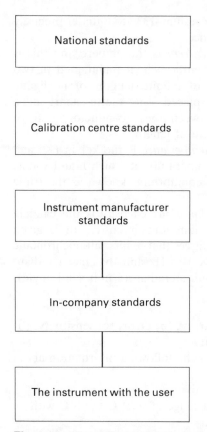

National standards

Calibration centre standards

Instrument manufacturer standards

In-company standards

The instrument with the user

Fig. 1.14 The calibration chain from national standards to an instrument in everyday use

Answer
The maximum error is −1.6°C.

Example 11
For the ammeter with a range 0–5 V giving the following calibration data, what is the accuracy of the meter as a percentage of full-scale-deflection?

Reading (V)	0	1	2	3	4	5
Deviation (V)	0	+0.02	+0.05	+0.01	0	−0.05

Answer
All the errors lie within plus or minus 0.05 V. This is a percentage of the full-scale-deflection of

$$\frac{0.05}{5} \times 100\%$$

Hence the accuracy is ± 1%.

Primary standards

The basic standards from which all others derive are the *primary standards*. There are primary standards for mass, length, time, current, temperature and luminous intensity. These are defined by international agreement and are maintained by national establishments, e.g., the National Physical Laboratory in Great Britain and the National Bureaux of Standards in the United States.

The primary standard of *mass* is an alloy cylinder (90% platinum, 10% iridium) of equal height and diameter, held at the International Bureau of Weights and Measures at Sèvres in France. The mass is defined as one kilogram. Duplicates of this standard are held in other countries.

The primary standard of *length* is the metre and is defined as the length of path travelled by light in a vacuum during a time interval of 1/299 792 458 of a second.

The primary standard of *time* is the second and this is defined as a duration of 9 192 631 770 periods of oscillation of the radiation emitted by the caesium–133 atom under precisely defined conditions of resonance.

The primary standard of *current* is the ampere and this is defined as that constant current which, if maintained in two straight parallel conductors of infinite length, of negligible circular cross-section, and placed one metre apart in a vacuum, would produce between these conductors a force equal to 2×10^{-7} N per metre of length.

The primary standard of *temperature* is the kelvin (K) and this is defined so that the temperature at which liquid water, water vapour and ice are in equilibrium (known as the triple point) is 273.16 K.

The primary standard of *luminous intensity* is the candela and this is defined as the luminous intensity, in a given direction, of a specified source that emits monochromatic radiation of frequency 540×10^{12} Hz and that has a radiant intensity of 1/683 watt per unit steradian (a unit solid angle).

Problems

1 Define the terms (a) accuracy, (b) error, (c) sensitivity, (d) repeatability, (e) reliability.
2 Explain the significance of the following information about instruments:
   (a) A balance has a sensitivity of 1 mg.
   (b) A thermometer has a range of $-20°C$ to $60°C$ with an accuracy of $\pm 0.3°C$.
   (c) An ammeter has a range of 0–30 A and an accuracy of $\pm 1\%$ f.s.d.
   (d) The total error due to non-linearity, hysteresis and non-repeatability for a load cell used to measure weight in the range 50 kg–1000 kg is $\pm 0.1\%$ of the reading indicated by the instrument.
3 A thermometer having an accuracy of $\pm 0.4°C$ gives a reading of 35.8°C. Within what temperature range will the actual temperature lie?
4 A voltmeter has a range of 0–10 V and an accuracy of $\pm 4\%$ of full-scale-deflection. Within what range of values will a voltage lie if the instrument gives a reading of 3.2 V?
5 A car speedometer gives no reading for speeds below 20 km/h and has a maximum possible reading of 160 km/h. What is (a) the dead space and (b) the range of the instrument?
6 A Bourdon pressure gauge with a range of 0–5000 kPa gave the following data in a calibration test. What is (a) the maximum error and (b) the accuracy as a percentage of the full-scale-deflection?

Gauge reading (kPa)	0	1000	2000	3000	4000	5000
Deviation (kPa)	0	−20	−20	−10	+10	+30

7 A voltmeter with a range of 0–12 V gave the following data in a calibration test. What is the accuracy of the instrument as a percentage of the full-scale-deflection?

Instrument reading (V) 0   2   4   6   8   10   12
Deviation (V)   0 −0.1 −0.2 −0.1 0 +0.1 +0.3

8 An iron–constantan thermocouple gives the following voltages at different temperatures. What will be the maximum non-linearity error over the temperature range if the thermocouple voltage is assumed to be directly proportional to the temperature?

Temperature °C 0   40   80   120   160   200
Voltage (mV)   0 2.06 4.19 6.36   8.56   10.78

9 Repeated measurements of the temperature of a furnace gave the following results. What is the average result and its accuracy?

Temperature °C 850 860 855 865 860 850 845

10 Explain the difference between random and systematic errors and explain how the effects of random errors can be eliminated.

2 Sensing elements

Introduction

The sensing element is the element in a measurement system that takes information about the condition being measured and transforms it into a more suitable form. The output from a sensing element can take many forms, e.g., a change in electrical resistance, a change in electrical capacitance, elastic deformation (perhaps it stretches) or an e.m.f.

> Information about the condition being measured
> $\longrightarrow$ information in a suitable form for measurement
> system

For example, the electrical resistance of a coil of wire depends on its temperature. Thus a resistance element can be used as a temperature-sensing element. Information about the temperature of a hot body is transformed into information in the form of resistance.

> Temperature changes $\longrightarrow$ resistance changes

A simple example of the use of a mechanical sensing element is a spring balance. The length of a spring depends on the forces used to stretch it. Thus a spring balance has as its input a force. Information about the force is changed into information in the form of a length change.

> Force change $\longrightarrow$ length change

Many measuring systems use more than one sensing element. Thus a measurement system used to determine the rate at which a liquid flows along a pipe might use firstly a sensing element which transforms information about the rate of flow into a pressure difference. A second sensing element might then be used to transform this pressure difference into a change in electrical resistance.

> Flow rate
> $\longrightarrow$ a pressure difference
> $\longrightarrow$ electrical resistance change

Resistive sensing elements

All these elements produce a change in electrical resistance. The input to the sensing element which produces this change might be a change in temperature, a change in strain, a change in position or a change in the intensity of light falling on it. The following gives more information about these forms of sensing elements.

Metal wire resistance thermometers

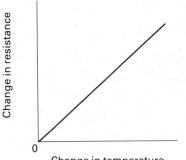

Fig. 2.1 Resistance–temperature graph for a metal

Figure 2.1 shows how the electrical resistance of a typical metal varies with temperature. The change in resistance of most metals is reasonably proportional to the change in temperature. For such a linear relationship (see Ch. 8 for a more detailed discussion):

Change in resistance is proportional to change in temperature θ.

Change in resistance $= R_0 \alpha \theta$

where R_0 is the resistance at $0°C$. α is called the temperature coefficient of resistance, unit $/°C$, and depends on the metal concerned. Platinum, nickel and copper are widely used for *resistance thermometers* (see Ch. 8). They have temperature coefficients of resistance of about:

platinum 0.0039 /°C
nickel   0.0067 /°C
copper   0.0038 /°C

As well as metals, *thermistors* are used for temperature measurement. Thermistors are small pieces of material made from mixtures of metal oxides, such as those of chromium, cobalt, iron, manganese and nickel. The material is formed into various forms of element, such as beads, discs and rods (Fig. 2.2). The resistance of thermistors generally decreases with an increase in temperature and is highly non-linear though there are some for which the resistance increases with an increase in temperature. Figure 2.3 shows a typical graph. The change in resistance per degree change in temperature is considerably larger than that which occurs with metals.

Both the metal wire and thermistor sensing elements have inputs of temperature changes and outputs of changes in resistance.

temperature change $\longrightarrow$ resistance change

Example 1

What is the change in resistance of a platinum resistance coil of resistance 100 Ω at $0°C$ when the temperature is raised to $30°C$? The temperature coefficient of resistance can be taken as 0.0039 /°C.

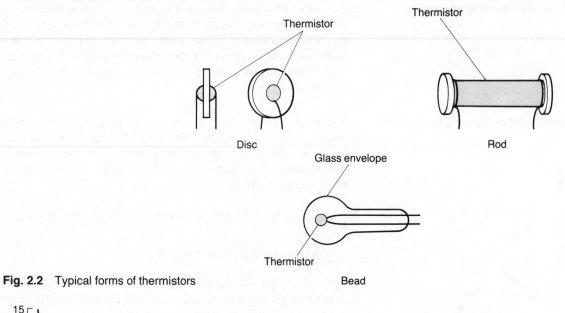

Fig. 2.2   Typical forms of thermistors

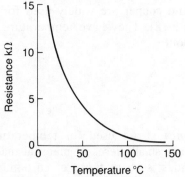

Fig. 2.3   Resistance–temperature graph for a thermistor

Answer

Assuming that the resistance varies linearly with temperature,

$$\text{Change in resistance} = R_0\alpha t$$

$$= 100 \times 0.0039 \times 30 = 11.7\ \Omega$$

Strain gauges

Generally when you stretch a strip of material its resistance increases. The more it is stretched the higher its resistance. The term *strain gauge* is used for a metal wire or foil resistance element or a semiconductor strip which is wafer-like and can be stuck onto surfaces like a postage stamp and whose resistance changes when subject to strain (Fig. 2.4). Strain is defined as being

$$\text{strain} = \frac{\text{change in length}}{\text{original length}}$$

The change in resistance is proportional to the strain. The change in resistance of a gauge of resistance R when subject to a strain ε is given by

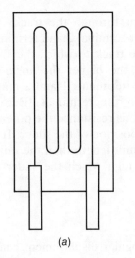

(a)

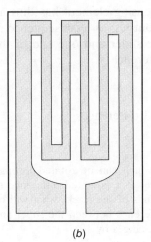

(b)

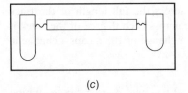

(c)

Fig. 2.4 Strain gauges, (a) metal wire, (b) metal foil, (c) semiconductor

Potentiometers

change in resistance $= RG\varepsilon$

where G is a constant called the gauge factor.

The wire strain gauge consists of a length of wire wound in a grid shape and attached to a suitable backing material (Fig. 2.4(a)). The metal foil strain gauge consists of a grid form which has been etched from a metal foil (Fig. 2.4(b)) and mounted on a resin film base. Typical values of gauge factor for metal wire and foil gauges are about 2.0.

Semiconductor strain gauges (Fig. 2.4(c)) are generally just strips of silicon doped with small amounts of p or n-type material. They have very high gauge factors, of the order of 100 to 175 for p-type silicon and -100 to -140 for n-type silicon. The negative gauge factor means that the resistance decreases with an increase in strain, unlike all the other forms of gauges where a positive gauge factor indicates an increase in resistance with an increase in strain. Semiconductors have a great advantage over metal gauges of a high gauge factor but do however have the disadvantage of being much more sensitive to changes of temperature.

Whatever the form of the strain gauge the input to such a sensing element is a change in strain and the output a change in resistance.

change in strain $\longrightarrow$ change in resistance

Example 2

A strain gauge has a resistance of 120 Ω and a gauge factor of 2.1. What will be the change in resistance produced if the gauge is subject to a strain of 0.0005?

Answer

Using the equation given above

Change in resistance $= RG\varepsilon = 120 \times 2.1 \times 0.0005 = 0.13$ Ω

The *rotary potentiometer* consists of a circular wire wound-track or a film of conductive plastic over which a rotatable electrical contact, the slider, can be rotated (Fig. 2.5). The track may be just a single turn or helical.

With a constant input voltage, between terminals 1 and 3 in

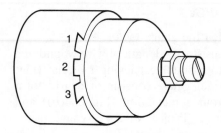

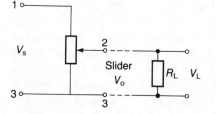

Fig. 2.5 Rotary potentiometer

Fig. 2.5 there is a constant current through the track. The output voltage V_o between the slider terminal 2 and terminal 1 depends on the position around the track to which the slider has been rotated. This is because the bigger the angle the greater the length of track between terminals 1 and 2 and so the greater the resistance between the terminals. Because there is a constant current through this resistance the potential difference across it depends on the position of the slider. If the track has a constant resistance per unit length then the output is proportional to the angle θ through which the slider has been rotated.

V_o is proportional to θ

$$V_o = k\theta$$

where k is a constant. Hence an angular displacement can be converted into a potential difference.

The rotary potentiometer as a sensing element has thus an input of a change in angle and an output, as a consequence of a change in resistance between the slider and one end of the track, of a change in potential difference.

Change in angle
$\longrightarrow$ change in resistance
$\longrightarrow$ change in potential difference

Example 3

A rotary potentiometer has a slider which can be rotated through 340° and a track of uniform resistance. If a potential difference of 6.0 V is connected between the two ends of the track, what will be the change in the output potential difference between the slider and one end of the track when the slider changes by 10°?

Answer

The potential difference of 6.0 V across the full length of the track means that the potential difference across each degree of the track is 6.0/340 = 0.0176 V/degree. Hence a change of 10° means a change in output of 0.176 V.

Photoconductive cells

A photoconductive cell (Fig. 2.6) has a resistance which depends on the intensity of light falling on it. Cadmium sulphide is a commonly used material because it has a response to the colours of the spectrum which is very similar to that of the human eye. The form of such a cell is a ceramic substrate on which there is a layer of photoconductive material in the form of a flat coil. The assembly is enclosed in a case with a glass window allowing light to fall on the photoconductive coil.

The photoconductive cell as a sensing element has thus an

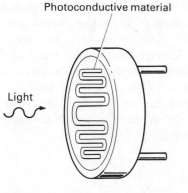

Photoconductive material

Light

Fig. 2.6 A photoconductive cell

Capacitive sensing elements

input of a change in light intensity and an output of a change in resistance.

Change in light intensity $\longrightarrow$ change in resistance

A capacitor is an electrical component which essentially consists of two plates separated by an insulator. When it is connected into an electrical circuit and a potential difference connected across it then one of the plates becomes positively charged and the other negatively charged. Increasing the potential difference increases the amount of charge on a plate. The term *capacitance* is defined as:

$$\text{capacitance} = \frac{\text{charge on a plate}}{\text{p.d. between the plates}}$$

A common form of capacitor consists of two parallel plates with an insulator between them. The capacitance depends on:

1 *The distance between the plates* The capacitance is doubled when the distance is halved. The capacitance is proportional to $1/d$, where d is the distance between the plates.
2 *The area of the plates* The capacitance is doubled if the areas of the plates are doubled. The capacitance is proportional to the area. The area we are referring to is the area of one plate which is directly opposite an identical area of the other plate. If the plates are not directly opposite each other then the area concerned is the area by which one plate overlaps the other.
3 *The material between the plates* The material between the plates is called the dielectric. A change in this material will produce a change in capacitance.

A change in capacitance can thus be produced by changing the separation of the plates, or the area of overlap, or the dielectric between the plates.

Variable plate separation sensing element

The capacitance of a parallel plate capacitor depends on the plate separation and so a change in this separation produces a

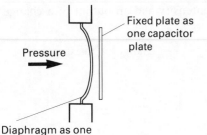

Pressure

Fixed plate as one capacitor plate

Diaphragm as one capacitor plate

Fig. 2.7 A capacitive pressure gauge

change in capacitance. Figure 2.7 shows a pressure gauge based on this principle. It consists of a circular diaphragm, held at the edges, acting as one plate of the capacitor and a fixed plate for the other one. Changes in pressure cause the diaphragm to distort and so change the separation between it and the fixed plate. The result is a change in capacitance.

The pressure gauge has thus an input of a change in pressure and, as a consequence of the deformation of the diaphragm, an output of a change in capacitance.

Change in pressure
$\longrightarrow$ deformation of diaphragm
$\longrightarrow$ change in capacitance

Variable plate area sensing element

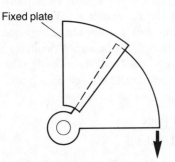

Fixed plate

Fig. 2.8 Variable plate area capacitive sensing element

The capacitance of a parallel plate capacitor depends on the overlap area of the two plates and so a change in area produces a change in capacitance. Figure 2.8 shows the form of a sensing element based on this principle. Such a sensing element has an input of a change in angle and an output of a change in capacitance.

Change in angle
$\longrightarrow$ change in overlap area
$\longrightarrow$ change in capacitance

Variable dielectric sensing element

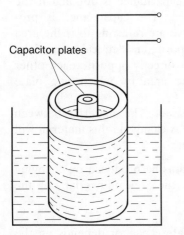

Capacitor plates

Fig. 2.9 Capacitive liquid level gauge

The capacitance of a parallel plate capacitor depends on the dielectric between the plates. Thus if the relative amount of two dielectrics between the plates varies then the capacitance varies. Figure 2.9 shows the basic form of such a sensing element for the determination of the level of a liquid. The liquid is an electrical insulator. The capacitor plates are two concentric cylinders. Between the lower part of these cylinders the dielectric is the liquid, between the upper parts air. With the liquid between the plates a higher capacitance is produced than when there is air. Thus changes in the level of the liquid change the capacitance.

Change in level
$\longrightarrow$ change in amount of liquid dielectric between plates
$\longrightarrow$ change in capacitance

Inductive sensing elements

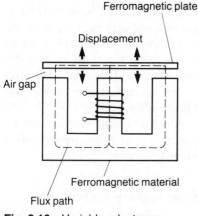

Fig. 2.10 Variable reluctance transducer

Variable reluctance

Variable differential inductor

If we took an electrical circuit of a battery connected to just a resistor and changed the resistance then the current throughout the circuit would change. In a similar way to this we have magnetic circuits. A simple magnetic circuit might consist of a just an iron ring (iron is referred to as a ferromagnetic material) with a coil of wire wrapped round part of it. When a current is passed through the coil it behaves like the battery in the electrical circuit. Instead, however, of producing an electric current in the circuit it produces magnetic flux in the circuit, i.e. the iron ring. We can change the electric current in a circuit by changing the electrical resistance, with the magnetic circuit we can change the magnetic flux by changing the *reluctance*. The electrical resistance of a piece of wire depends on its length, its cross-sectional area and the material of the wire (the term *conductivity* being used). The reluctance of a piece of material depends, just like electrical resistance, on its length, its cross-sectional area and the material (the term *permeability* being used).

Figure 2.10 shows the basis of a variable reluctance sensing element. Air has a much lower permeability than a ferromagnetic material. This means that introducing an air gap into a magnetic circuit is like introducing a high resistance into an electrical circuit. The bigger the air gap the greater its reluctance. This means that changing the size of the air gap changes the reluctance and hence changes the magnetic flux in the magnetic circuit. The air gap is changed by moving the ferromagnetic plate.

Thus a variable reluctance sensing element has an input of a displacement and an output of a change in reluctance.

change in displacement ⟶ change in reluctance

When a current passes through a coil it produces a magnetic field around and through the coil. This magnetic field is constant when the current is constant. However a changing current will produce a changing magnetic field. Now when a coil is in a changing magnetic field electromagnetic induction occurs and an e.m.f. is induced in the coil. Thus when the current through a coil is changing the resulting changing magnetic field means that an e.m.f. is induced in the coil. The size of the induced e.m.f. depends on:

- the rate at which the current changes;
- the size of the coil;
- the material inside the coil.

The factors relating to the size of the coil and the material inside it can be described by the term *inductance*. Thus the size of the induced e.m.f. depends on the rate of change of current and the coil inductance.

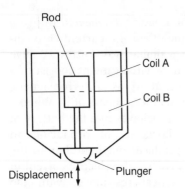

Fig. 2.11 Variable differential inductor

Variable differential transformer

Figure 2.11 shows the basic elements of a *variable differential inductor*. It consists of two coils between which a core rod is moved. The inductance of the coil depends on what material is in its core. Thus the movement of a rod into the core of a coil can, for a suitable rod material, have a marked effect on its inductance. When the rod has the same length in each of the two coils they have the same inductance. Movement of the rod from this position then results in the inductance of one coil increasing and the other coil decreasing.

Thus a variable differential inductor sensing element has an input of a change in displacement and an output of a difference in inductance between two coils.

Change in displacement
$\longrightarrow$ difference in inductance between two coils

A current through a coil produces a magnetic field in the space around the coil. When there is a changing current in a coil a changing magnetic field is produced. If there is another coil in the vicinity then this changing magnetic field can result in an induced e.m.f. being produced in that coil. This is the basis of the transformer. With the transformer an alternating current through one coil results in an alternating magnetic field and the induction of an alternating e.m.f. in a second coil. The size of the alternating e.m.f. in this second coil depends on the number of turns of the coil and the size of the magnetic field which interacts with its turns. This magnetic field is affected by the presence of an iron core passing through both the coils.

The *linear variable differential transformer*, generally referred to by the abbreviation *LVDT*, consists of a transformer with a primary coil and two secondary coils, as in Fig. 2.12. When there is an alternating voltage input to the primary coil alternating e.m.f.s are induced in the secondary coils. Both the secondary coils are identical and thus with the core central and so equal amounts in each coil, the e.m.f.s induced in the two secondary coils will be the same. The two secondary coils are connected so that their outputs oppose each other. Thus when both give the same output the net result is zero output. However, when the core is displaced from the central position there is a greater amount in one coil than the other. The result of this displacement is that the induced e.m.f. in one coil is greater than that in the other. The output, the difference between the two e.m.f.s, is thus a measure of the displacement of the core from its central position.

The linear variable differential transformer thus has an input of a displacement and an output of an alternating voltage, the size of the voltage being related to the displacement.

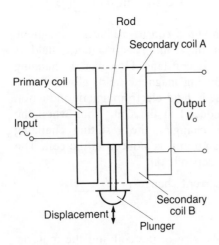

Fig. 2.12 Linear variable differential transformer

change in displacement $\longrightarrow$ change in size of a.c. voltage

Thermoelectric sensing elements

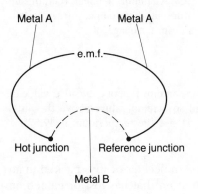

Fig. 2.13   A thermocouple

If two different metals are joined together a potential difference occurs across the junction. This potential difference depends on the metals used and the temperature of the junction. A *thermocouple* is a complete circuit involving two such junctions (Fig. 2.13). If both junctions are at the same temperature then the potential differences across each junction are the same but in opposite directions and so there is no net e.m.f.. If however there is a difference in temperature between the two junctions then there is an e.m.f., this being called the *thermoelectric e.m.f.*. The value of this e.m.f. depends on the two metals concerned and the temperatures of both the junctions. Figure 2.14 shows, for a number of pairs of metals, how the thermoelectric e.m.f. varies with temperature when one junction is held at 0°C. Standard tables are available for the metals usually used for thermocouples.

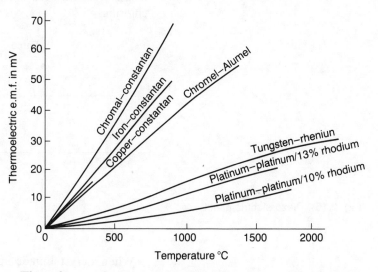

Fig. 2.14   Thermoelectric e.m.f. temperature graphs

Thus for a thermocouple the input is a difference in temperature between its two junctions and the output is an e.m.f.

Temperature difference $\longrightarrow$ e.m.f.

Example 4

Using the data given below for a chromel–constantan thermocouple, estimate the temperature of one junction when the other is at 0°C and the thermoelectric e.m.f. is 3.020 mV.

Temp. °C	0	10	20	30	40	50
e.m.f. mV	0.0000	0.591	1.192	1.801	2.419	3.047

Answer

The relationship between the e.m.f. and the temperature is virtually linear, rising by about 0.06 mV/°C. The 3.020 mV e.m.f. indicates a temperature between 40 and 50°C. Assuming a linear relationship between the e.m.f. and temperature in that temperature interval, then a 1°C change would mean an e.m.f. change of

$$\frac{3.047 - 2.419}{10} = 0.0628 \text{ mV}$$

Thus the e.m.f. of 3.020 mV means an e.m.f. above the 40°C value of (3.020 − 2.419) = 0.601 mV and so a temperature above the 40°C value of 0.601/0.0628 = 9.57°C. Hence the temperature is 49.57°C.

Piezo-electric sensing elements

Crystals are made up of atoms, molecules or ions packed in an orderly manner. An ion is an atom that has lost or gained an electron and so has either a positive or a negative charge. Thus, for example, a sodium chloride crystal, common salt, consists of positive ions of sodium and negative ions of chlorine.

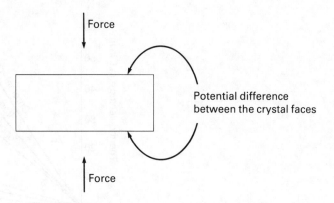

Fig. 2.15 Piezo-electric effect

When a crystal made up of ions is stretched or squashed the ions in the crystal are displaced from their normal positions. This results in a potential difference appearing across the crystal (Fig. 2.15). The effect is called *piezo-electricity*. Suitable crystals are quartz, tourmaline and what are called piezo-electric ceramics, such as lead zirconate-titanate. The input for a piezo-electric sensing element is thus forces which squash the crystal and the output a potential difference.

Forces squashing a crystal $\longrightarrow$ potential difference

The piezo-electric effect can operate in reverse, i.e., a potential difference applied to a crystal can result in it expanding or contracting. If an alternating potential difference is used the crystal surfaces move back-and-forth at the frequency of the potential difference. At high frequencies this oscillatory movement of the crystal faces produces a pressure wave in the surrounding medium which is referred to as an *ultrasonic wave*.

Alternating potential difference
$\longrightarrow$ alternating movement of crystal faces
$\longrightarrow$ ultrasonic wave

The existence of an ultrasonic wave can be detected by it impinging on a piezo-electric crystal. The pressure wave will alternately squash and expand the crystal and so produce an alternating potential difference which can be detected.

Ultrasonic wave
$\longrightarrow$ squashing and expanding a crystal
$\longrightarrow$ alternating potential difference

Elastic sensing elements

The *spring balance* is a widely used device for the measurement of forces. It depends on the principle of the forces producing changes in the length of a spring. The spring is just one example of what is called an elastic sensing element. Other forms include elastic deformation of rings, cylinders, diaphragms, capsules, bellows and tubes. All of them are based on the principle that changes in forces produce changes in shape or length.

Proving ring

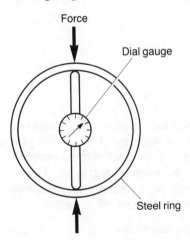

Force

Dial gauge

Steel ring

Fig. 2.16 Proving ring

Figure 2.16 shows a steel ring, called a *proving ring*. The action of forces is to deform the ring, the amount of deformation being a measure of the force. This deformation can be measured by a dial test indicator gauge, as shown in Fig. 2.16 or by strain gauges (see earlier in this chapter) which are attached to the ring. The proving ring has thus an input of forces and an output of a change in shape which is then sensed by means of another sensor.

Change in forces
$\longrightarrow$ change in shape
$\longrightarrow$ some other change,
e.g., change in strain of strain gauges

Load cells

A cylinder will deform under the action of forces (Fig. 2.17), its walls bowing out when the forces squash the cylinder. This deformation of the cylinder walls can be measured by means of strain gauges. Such an arrangement is one form of what is termed a *load cell* since it can be inserted in a structure, perhaps the legs supporting some container, and used to determine the load. A wide variety of forms are used for load cells, the cylinder being just one possibility. The load cell, whatever its form, has an input of forces and an output of a change in shape. This change in shape is then sensed by means of another sensor, e.g. strain gauges.

Change in forces
$\longrightarrow$ change in shape
$\longrightarrow$ some other change,
e.g., change in strain of strain gauges

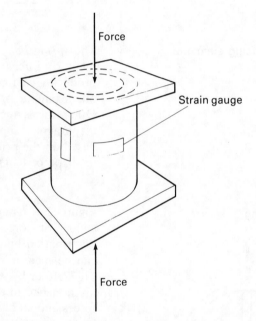

Fig. 2.17 Load cell

Diaphragms, capsules and bellows

A diaphragm is just the term used for a disc clamped round its edges. It will bow in or out at its centre under the action of a force acting at right-angles to its surface. Such a force occurs if there is a difference in pressure between its two sides. Thus a diaphragm can be used as a sensor, with an input of a pressure difference and an output of a displacement of its centre.

Pressure difference $\longrightarrow$ displacement

The amount of movement with a plane diaphragm (Fig. 2.18(*a*)) is fairly limited, however greater movement is possible with corrugations in the diaphragm (Fig. 2.18(*b*)). Even greater movement is possible if two corrugated diaphragms are combined to give a capsule (Fig. 2.18(*c*)). A stack of capsules is just a bellows arrangement (Fig. 2.18(*d*)) and

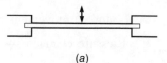

(a)
Plane diaphragm

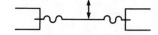

(b)
Corrugated diaphragm

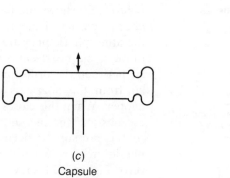

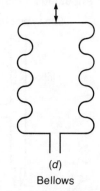

Fig. 2.18 Diaphragms, capsules and bellows

(c)
Capsule

(d)
Bellows

this is even more sensitive. Diaphragms, capsules and bellows are made from such materials as stainless steel, phosphor bronze, and nickel, with rubber and nylon also being used for some diaphragms.

Bourdon tubes

There is a toy which consists of a rolled up paper tube, open at one end and closed at the other. When you blow into the tube the roll opens out. This is the basis of a sensing element called a *Bourdon tube*. Figure 2.19 shows two versions of such a tube. When the pressure inside the C-shaped tube (Fig. 2.19(a)) is increased it opens up to some extent. The amount by which it opens out is a measure of the pressure. Thus the result of a pressure change in the tube is a straightening out

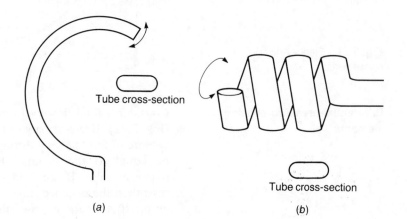

Tube cross-section

Tube cross-section

Fig. 2.19 Bourdon tubes

(a)

(b)

and hence a movement of the free end of the tube.

Pressure change $\longrightarrow$ straightening out

A helical form of such a tube (Fig. 2.19(*b*)) gives a greater deflection. The tubes are made from such materials as stainless steel and phosphor bronze.

Pneumatic sensing elements

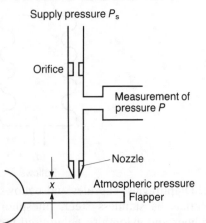

Fig. 2.20 Flapper-nozzle

Figure 2.20 shows the basic form of what is termed a *flapper–nozzle* sensing element. Air at a constant pressure P_s above the atmospheric pressure flows through the orifice and escapes through the nozzle into the atmosphere. The pressure of the air between the orifice and nozzle is measured. The escape of air from the nozzle is controlled by the movement of the flapper. When this closes off the nozzle, i.e. $x = 0$, then no air escapes and the measured pressure equals P_s. As x increases so the pressure P decreases, becoming equal to the atmospheric pressure when x is very large, i.e. a gauge pressure of zero. Figure 2.21 shows how the measured gauge pressure P varies with the displacement x of the flapper. The measured pressure can thus be used as a measure of the displacement x.

change in displacement $\longrightarrow$ change in pressure

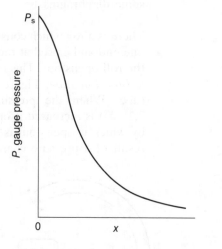

Fig. 2.21 Flapper-nozzle characteristic

Differential-pressure sensing elements

Consider a fluid flowing from a wide pipe to a narrower pipe (Fig. 2.22). If the density of the fluid does not change then the volume of fluid flowing through the wide tube per second must be equal to the volume flowing per second through the narrower tube. If the fluid has a velocity of v_1 in the wider tube then the distance travelled by the fluid in one second will be v_1. If A_1 is the cross-sectional area of the wider tube then the volume passing through the wider tube in one second is A_1v_1. Similarly for the narrower tube, the volume passing through it per second will be A_2v_2. Hence

$$\text{Volume passing/second} = A_1v_1 = A_2v_2$$

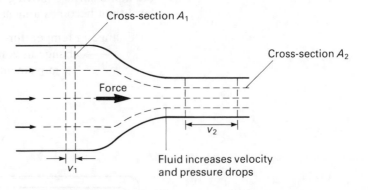

Fig. 2.22 Fluid flow through a constriction

This relationship is known as the *continuity equation*. Consequently, when the tube narrows and the cross-sectional area decreases then the velocity must increase. The fluid velocity in the narrower tube is greater than that in the wider tube. The fluid in moving from the wider to the narrower tube must therefore accelerate. This means there is a force pushing the liquid in that direction. This force arises from a pressure difference between the fluid in the wider and narrower tubes. The pressure in the wider tube is greater than that in the narrower tube. Thus there is a pressure drop when the fluid velocity increases. The size of the pressure drop depends on the rate of flow of the fluid through the pipe.

rate of flow is proportional to $\sqrt{}$(pressure drop)

Hence the rate of flow can be determined by a measurement of the pressure drop (see Ch. 7 for more details). The input to such a sensor is flow rate and the output is a pressure difference.

Flow rate $\longrightarrow$ pressure difference

Expansion sensing elements
Bimetallic strip

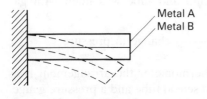

Fig. 2.23 Bimetallic strip

When the temperature of a strip of metal increases it expands. The *bimetallic strip* consists of strips of two different metals which are attached to each other at their ends (Fig. 2.23). One of the metals has a greater expansivity than the other. Thus when the temperature rises, the only way one of the metals can expand more than the other is for the strip to curve. In Fig. 2.23 metal A has expanded more than metal B. For a particular pair of metals the amount by which the bimetallic strip curves depends on the temperature change.

Change in temperature $\longrightarrow$ change in curvature

Liquid expansion

When the temperature of a liquid is increased it increases in volume, i.e., expands. Use is made of this for the measurement of temperature in the *liquid-in-glass* thermometer. With the thermometer the increase in the volume of the liquid results in the liquid surface moving up a tube. The position of the liquid level then becomes a measure of the temperature.

> Change in temperature
> $\longrightarrow$ change in volume
> $\longrightarrow$ movement of liquid surface up a tube

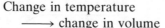

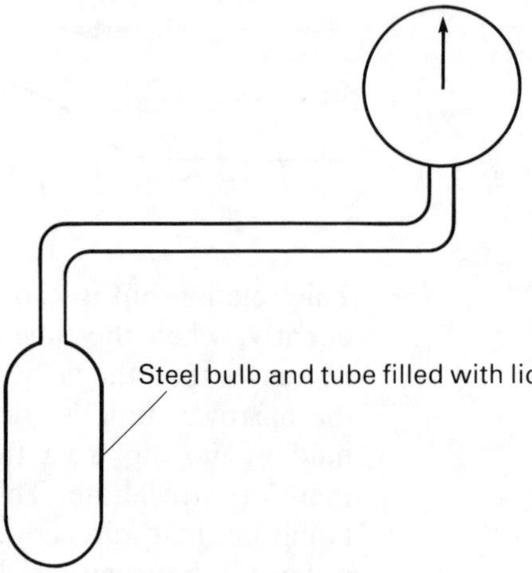

Pressure gauge

Steel bulb and tube filled with liquid

Fig. 2.24 Liquid-in-steel thermometer

Another form of thermometer which relies on liquid expansion is the *liquid-in-steel* thermometer. This consists of a steel bulb connected via a fine steel tube to a pressure gauge (Fig. 2.24). The arrangement contains a liquid. When the temperature rises the liquid expands and so the pressure above the liquid increases. This pressure is measured using a pressure gauge, e.g. a Bourdon tube (see earlier this chapter) and is a measure of the temperature.

> Change in temperature $\longrightarrow$ change in pressure

Gas expansion

If a gas is not allowed to increase in volume when the temperature increases then its pressure increases. This change in pressure is a measure of the change in temperature. An industrial form of such a thermometer would be similar to that of the liquid-in-steel thermometer shown in Fig. 2.24, the only difference being that the bulb and tube was filled with gas rather than liquid.

> Change in temperature $\longrightarrow$ change in pressure

Evaporation

With the *vapour pressure* thermometer the arrangement is as described in Fig. 2.24, i.e., a sealed tube and a pressure gauge, but the tube is partially filled with a volatile liquid. When the temperature rises evaporation takes place from the liquid and the pressure due to the vapour above the liquid surface increases. This vapour pressure is then a measure of the temperature.

Change in temperature $\longrightarrow$ change in vapour pressure

Problems

1 Sensors change information about a quantity into information in another form. For the following sensors, what information changes are occurring?
(a) Resistance thermometer element
(b) Thermocouple
(c) Photoconductive cell
(d) Strain gauge
(e) Bourdon tube
(f) Mercury-in-glass thermometer

2 Specify a sensor, or primary and secondary sensors, which could be used for each of the following information-change situations.
(a) Displacement to potential difference
(b) Force to displacement
(c) Force to resistance change
(d) Temperature to resistance
(e) Temperature to e.m.f.
(f) Temperature to pressure

3 State the principles involved in the following sensing elements.
(a) A resistance strain gauge
(b) A bellows pressure sensor
(c) A Bourdon tube
(d) A linear variable differential transformer
(e) A piezo-electric ultrasonic wave detector
(f) A vapour pressure thermometer

3 Signal converters

The signal from the sensing element of a measurement system is likely to need some further processing before it can be used to give a visual display. This may be because the signal is:

1 *In an inconvenient form* For example, a resistance thermometer sensing element gives a change in resistance when the temperature changes. This might need to be converted into a change in current or potential difference so that a reading can be obtained on a meter.

2 *Too small or perhaps has too much noise associated with it* Thus an amplifier might be used to increase the size of the signal and filters to remove noise. The term *noise* is used for signals that are detected which are not the ones wanted. Thus an electrical instrument might pick up noise from the mains electricity supply.

3 *Not in the right form for transmission over a distance* The sensing element might be in a location which is some distance from the console where the display is located and thus might need to be changed in form so that it can be transmitted over the distance.

The term *signal conditioning* is generally used for the element or elements in a measurement system which convert the signal from the sensing element into a form suitable for the display unit. Typical signal conditioning elements are bridges where a change in resistance, capacitance or inductance can be converted to a change in potential difference.

The term *signal processing* is used for the element or elements which are concerned with improving the quality of the signal. This involves such processes as signal amplification and signal filtering.

The term *signal transmission* is used for the element or elements which are used for conveying the signal from the sensing element over some distance to the display.

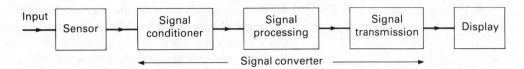

Fig. 3.1 A measurement system

Thus a measurement system containing all three of the above elements would be of the form shown in Fig. 3.1.

Signal conditioning

Signal conditioning might, for example, involve converting:

1 *A resistance change to a current or potential difference change* An example of this is the Wheatstone bridge (see p. 38) which can be used to convert the resistance change of a resistance thermometer element or a strain gauge into a current or potential difference which can then give an indication on a meter.

2 *An inductance or capacitance change into a current or potential difference change* An alternating current bridge might be used for this.

3 *A displacement change into a current or potential difference change* This might be the movement of a diaphragm because of a difference of pressure between its two faces. This movement can be converted into an electrical signal if strain gauges are subject to strain by the movement of the diaphragm (Fig. 3.2). A change in displacement of the diaphragm then results in changes in resistance of the strain gauges. If the strain gauges are in a Wheatstone bridge then the change in resistance can result in a change in current or potential difference.

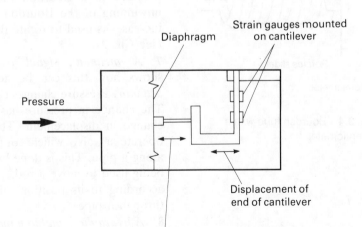

Fig. 3.2 A strain gauge pressure cell

4 *An electrical current into a mechanical rotation* The moving-coil galvanometer (Fig. 3.3) is an obvious example of this change. A current through the coil of the meter causes it to rotate and move a pointer across a scale.

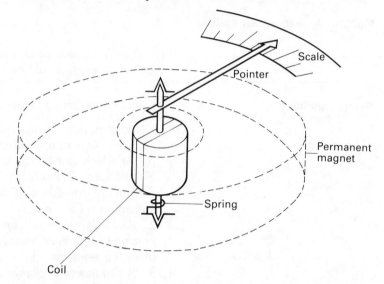

Fig. 3.3 The basic moving-coil galvanometer

5 *An electrical potential difference into a displacement* An example of this is the potentiometer measurement system in which potential differences or e.m.f.s are converted into movements of a potentiometer slider along a scale (see later this chapter).

6 *A pneumatic signal into an electrical signal* Figure 3.4 shows one possibility for converting a change in the air pressure in a Bourdon tube into an electrical signal. The unwinding of the Bourdon tube when the pressure inside it increases is used to rotate the slider of a rotary potentiometer (see Ch. 2).

7 *A pneumatic signal to a mechanical signal* Figure 3.5 shows how this can be achieved with what is termed an *actuator*. Pressure changes cause a rubber diaphragm to move. The change in pressure has thus become transformed into a change in displacement. The actuator may then be used to operate a valve which controls the rate of flow of a liquid along a pipe. This is done by the movement of the diaphragm being used to move a rod, which in turn moves a plug which, according to its position, allows more or less liquid to pass through a pipe.

8 *A hydraulic signal to a mechanical signal* Figure 3.6 shows an example of this conversion using a *hydraulic cylinder*. With the double-acting cylinder a pressure difference between the two sides of the piston causes it to move. With the single-

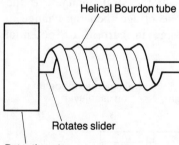

Fig. 3.4 Bourdon tube with a potentiometer

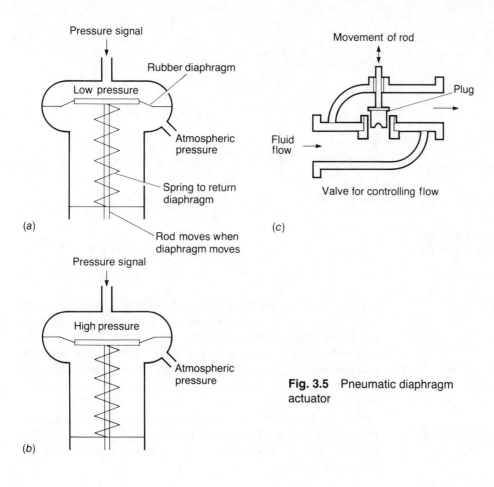

Pressure signal

Rubber diaphragm

Low pressure

Atmospheric pressure

Spring to return diaphragm

(a)

Rod moves when diaphragm moves

Pressure signal

High pressure

Atmospheric pressure

(b)

Movement of rod

Plug

Fluid flow

Valve for controlling flow

(c)

Fig. 3.5 Pneumatic diaphragm actuator

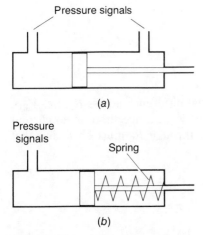

Pressure signals

(a)

Pressure signals

Spring

(b)

Fig. 3.6 Hydraulic cylinders, (a) double acting, (b) single acting with spring return

acting cylinder a pressure change on just one side of a piston causes it to move against the action of a spring. Thus a pressure change has become converted into a displacement change. The hydraulic cylinder can also be used with pneumatic signals.

9 *A mechanical signal to an optical signal* An example of this is the twisting or rotation of a wire or shaft which can be converted into the movement of a beam of light. A small mirror is attached to the wire or shaft and a beam of light directed at it. When the mirror is rotated the beam of light reflected from the mirror rotates (Fig. 3.7). Quite a small angle of rotation of the mirror can result in a large movement of the reflected beam of light over a scale placed some distance from it. This principle is used in ultraviolet recorders (see Ch. 4).

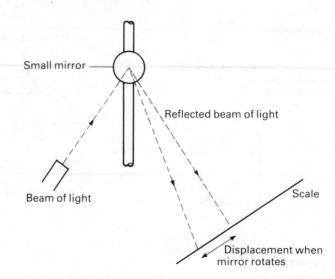

Fig. 3.7 Converting a mechanical rotation into movement of a beam of light

Wheatstone bridge

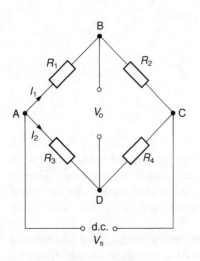

Fig. 3.8 The Wheatstone bridge

Figure 3.8 shows the basic form of the *Wheatstone bridge*. It has a d.c. supply and each of the four bridge arms is a resistance. The resistances in the arms of the bridge can be adjusted so that the output potential difference is zero. A galvanometer connected between the output terminals will then indicate zero current. In such a condition the bridge is said to be balanced.

When the output potential difference is zero then the potential at B must equal that at D. This means that the potential difference across R_1, i.e. V_{AB}, must equal that across R_3, i.e. V_{AD}. But

$$V_{AB} = I_1 R_1$$

and

$$V_{AD} = I_2 R_3$$

Thus

$$I_1 R_1 = I_2 R_3$$

It also means that the potential difference across R_2, i.e. V_{BC}, must equal that across R_4, i.e. V_{DC}. Since there is no current through BD then the current through R_2 must be I_1 and that through R_4 I_2. Thus

$$V_{BC} = I_1 R_2$$

and

$$V_{DC} = I_2 R_4$$

Hence

$$I_1 R_2 = I_2 R_4$$

Thus using our earlier equation

$$I_1R_1 = I_2R_3 = (I_1R_2/R_4)R_3$$

and so

$$\frac{R_1}{R_2} = \frac{R_3}{R_4}$$

This is the balance condition. It is independent of the supply voltage, depending only on the resistances in the four bridge arms. If R_2 and R_4 are known fixed resistances and R_1 the sensing element then R_3 can be adjusted to give the zero potential difference condition and R_1 determined from a knowledge of the values of R_2, R_3 and R_4. By a suitable choice of the ratio R_2/R_4 a small resistance change in R_1 can be determined by means of a much larger resistance change in R_3.

Suppose we now consider a balanced bridge and what happens when the resistance in one of the bridge arms (R_1) changes. The result of such a change will be to produce an output potential difference, i.e. a potential difference between points B and D. The change in the output V_o depends on the change in R_1. If the change in R_1 is much smaller than the values of the resistors then:

$$\text{change in output} = \frac{V_s}{R_1 + R_2} \times \text{change in } R_1$$

Under such conditions the change in output potential difference is proportional to the change in resistance of R_1. We can then consider the Wheatstone bridge to be a device which has an input of a change in resistance of a sensor and an output which is a change in potential difference (Fig. 3.9).

Input	Wheatstone bridge	Output
Change in resistance		Change in potential difference

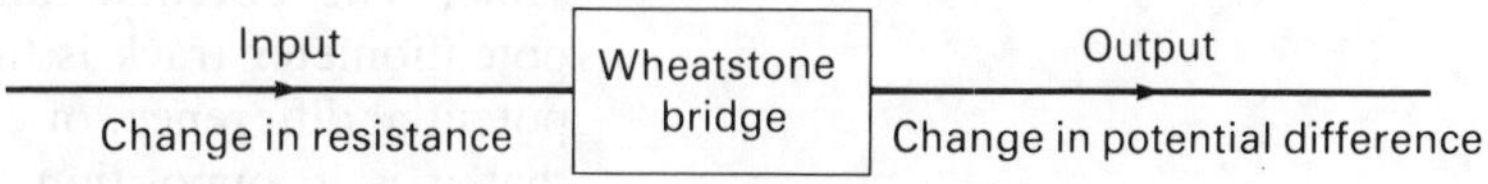

Fig. 3.9 The Wheatstone bridge signal conditioner

Example 1

A Wheatstone bridge has a resistance ratio of 1/100 for R_2/R_4, and R_3 is adjusted to give zero current. Initially this occurs with R_3 1000.3 Ω. The resistance R_1 then changes as a result of a temperature change and zero current is then obtained when R_3 is 1002.1 Ω. What was the change in resistance of R_1?

Answer

Initially

$$R_1 = \frac{R_2R_3}{R_4} = \frac{1}{100} \times 1000.3$$

After the change

$$R_1 + \text{change in } R_1 = \frac{1}{100} \times 1002.1$$

$$\text{change in } R_1 = \frac{1}{100} \times (1002.1 - 1000.3) = 0.018 \ \Omega$$

Example 2

A platinum resistance thermometer element has a resistance of 100 Ω at 20°C and is one of the arms of a Wheatstone bridge. The bridge is balanced at this temperature, all the other arms also having a resistance of 100 Ω. What will be the out-of-balance potential difference when the temperature of the platinum resistance thermometer element changes by 10°C and causes its resistance to change by 3.9 Ω? The bridge has a supply voltage of 6.0 V.

Answer

Using the equation given earlier

$$\text{change in output} = \frac{V_s}{R_1 + R_2} \times \text{change in } R_1$$

$$= \frac{6.0 \times 3.9}{100 \times 100} = 0.117 \text{ V}$$

Potentiometer measurement system

Figure 3.10 shows the basic form of a potentiometer measurement system. The working battery is used to produce a potential difference across the full length of the potentiometer track. If the track has a constant resistance per unit length then the potential difference across each unit of length is the same. The potential difference across the length L of the potentiometer track is thus proportional to the length L. The potential difference, or e.m.f., being measured is so connected that it is in opposition to the potential difference across the

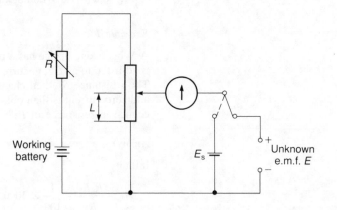

Fig. 3.10 Potentiometric measuring system

potentiometer track. The potentiometer slider is moved along the potentiometer track until no current is detected by the galvanometer. When this occurs the unknown e.m.f. E must be equal to the potential difference across the length L of the track. If the track is uniform then

$$E = kL$$

where k is a constant, in fact the potential difference per unit length of track. This can be determined by repeating the balancing operation with a standard cell of e.m.f. E_s. Then

$$E_s = kL_s$$

where L_s is the balance length with the standard cell. Hence

$$E = \frac{E_s L}{L_s}$$

With a commercial form of the potentiometer measuring system the movement of the potentiometer slider over the track results in the movement of a pointer over a scale. This scale is calibrated directly in volts. The standardisation is achieved by setting the pointer to the required value of the standard cell e.m.f. and then adjusting R until balance occurs. This change in resistance R changes the potential difference applied across the potentiometer track.

The potentiometer measurement system does not depend on the calibration of a galvanometer since it is only used to indicate when there is zero current. Because at balance no current is taken from the source being measured, then when a battery is measured the measurement gives the e.m.f. Because no current is taken, the arrangement is essentially a voltmeter of infinite resistance.

Example 3

A standard cell of e.m.f. 1.018 V gives a zero current reading with a potentiometer when the slider is at the mid-point of its track. When the potentiometer is used to measure an unknown potential difference the zero current reading is obtained when the slider is a quarter way along the track. What is the unknown potential difference?

Answer

If the track is assumed to have a constant resistance per unit length then the result with the standard cell indicates that the working battery is applying a potential difference of 1.018 V over half the track length and so 2.036 V over the full track length. Thus the unknown potential difference must be a quarter of 2.036 V, i.e. 0.509 V.

Signal processing

Signal processing might, for example, involve:

1 *Signal amplification* This might be the use of an electronic amplifier to make a small electrical signal bigger or a lever to make a mechanical displacement bigger (see later this chapter). Whatever the form a *transfer function* or *gain* can be defined for the amplifier as

$$\text{transfer function} = \frac{\text{output}}{\text{input}}$$

2 *Signal attenuation* Attenuation is used to describe the process of reducing the size of a signal.

3 *Signal filtering* This is to remove unwanted signals.

4 *Changing an analogue signal into a digital signal, or a digital signal into an analogue signal* An analogue signal is one in which the information being transmitted is in the form of a variation in size or strength of the signal, whereas a digital signal is one in which the information is transmitted as just a number, a set of digits. Most sensing elements give analogue signal outputs. If these outputs are to be processed by a computer then they need to be converted into digital signals.

Mechanical amplifiers

The *lever* is an example of a mechanical amplifier which can be used to change the size of a displacement signal from a sensing element.

$$\text{Small displacement} \longrightarrow \text{bigger displacement}$$

The transfer function of the lever, i.e. how much bigger the output is than the input, depends on the relative distances from the lever pivot point of the application of the input to the lever and of the extraction of the output. As indicated in Fig. 3.11, because of similar triangles

$$\frac{\text{input displacement}}{\text{input–pivot distance}} = \frac{\text{output displacement}}{\text{output–pivot distance}}$$

Hence

$$\text{transfer function} = \frac{\text{output}}{\text{input}} = \frac{\text{output–pivot distance}}{\text{input–pivot distance}}$$

Thus to obtain magnification of the input displacement, the input–pivot distance must be smaller than the output–pivot distance.

An instrument *pointer* is a simple example of a lever (Fig. 3.12). The pointer rotates about an axis. The input is at a distance r from this axis, where r is the radius of the shaft which transmits the rotation to the pointer. The output from

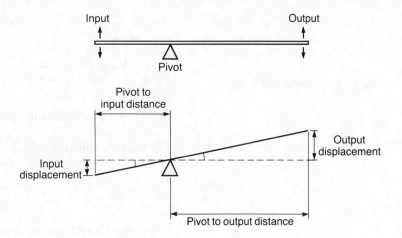

Fig. 3.11 The lever as a signal conditioner

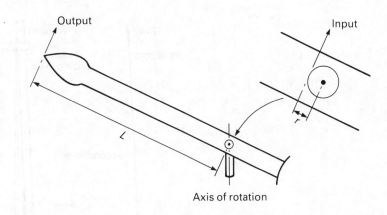

Fig. 3.12 Pointer

the pointer is its movement over the scale and this is a distance L from the axis. Thus the transfer function is L/r. This is how much the input has been multiplied to give the output.

Where a large magnification is required a *compound lever* may be used. This involves the output from the first lever becoming the input to a second lever (Fig. 3.13). For a two-lever system,

$$\text{transfer function of 1st lever} = \frac{\text{1st lever output}}{\text{input to levers}}$$

$$\text{transfer function of 2nd lever} = \frac{\text{output from levers}}{\text{1st lever output}}$$

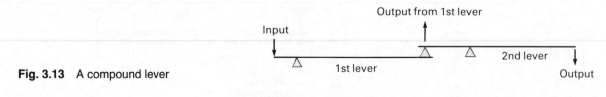

Fig. 3.13 A compound lever

The overall transfer function for the system of levers is

$$\text{transfer function} = \frac{\text{output from levers}}{\text{input to levers}}$$

$$= \frac{\text{output from levers}}{\text{1st lever output}} \times \frac{\text{1st lever output}}{\text{input to levers}}$$

Thus the overall transfer function is the product of the transfer functions of the two levers. Thus if the first lever has a transfer function of 10 and the second lever 12, then the output displacement will be $10 \times 12 = 120$ times bigger than the input displacement. Compound levers enable quite large amplifications of displacements to be obtained. Figure 3.14 shows an application of this in the *Huggenberger extensometer*.

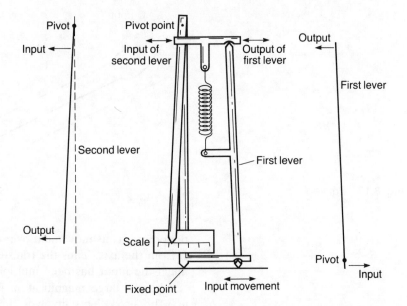

Fig. 3.14 The Huggenberger extensometer

Another way of producing amplification is by the use of *gear trains*. These have a rotation as an input and give a rotation as an output.

Small rotation $\longrightarrow$ bigger rotation

Figure 3.15 shows a simple gear train involving an input gear wheel which is rotated by one shaft and an output gear wheel which results in rotation of another shaft. Each tooth on the

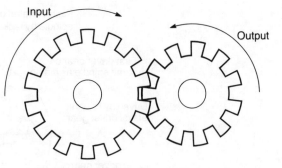

Fig. 3.15 A simple gear train signal conditioner

input gear wheel fits into a corresponding space between teeth on the output gear wheel. If there are N_I teeth on the input wheel and N_O on the output wheel then one complete revolution of the input shaft means that the output shaft rotates by the fraction (N_I/N_O) of a revolution. Thus the transfer function is

$$\text{transfer function} = \frac{\text{number of teeth on input gear } N_I}{\text{number of teeth on output gear } N_O}$$

Figure 3.16 shows an example of an instrument, a *Bourdon pressure gauge*, employing gears. A change in pressure in the Bourdon tube (see Ch. 2) causes the tube to straighten out to some extent. This movement causes a segment of a gear wheel to rotate and hence the wheel to which the pointer is attached.

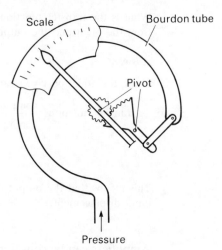

Fig. 3.16 Bourdon pressure gauge

Compound gear trains can be used to give greater magnifications. Figure 3.17 shows the basic form of the gears used in a *dial indicator gauge*. The linear movement of the plunger is translated into a rotary motion by means of the rack and pinion. This causes the first gear wheel to rotate and so as a consequence cause the second gear wheel to rotate. The result

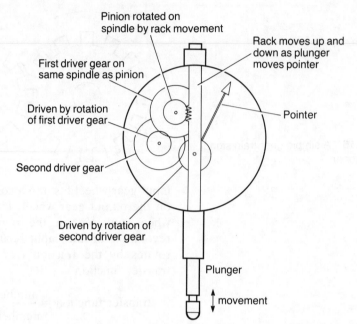

Fig. 3.17 The dial indicator gauge

of this is motion of the gear wheel on which the pointer is mounted and hence the movement of the pointer across the scale.

Example 4

What is the transfer function of a lever for which the input is 4 mm from the pivot and the output 40 mm?

Answer

Using the equation given earlier

$$\text{transfer function} = \frac{\text{output}}{\text{input}} = \frac{\text{output–pivot distance}}{\text{input–pivot distance}}$$

$$= \frac{40}{4} = 10$$

This means that the output displacement is 10 times bigger than the input displacement.

Electronic amplifier

Amplifiers can be considered to have as input a small potential difference and give as output a bigger potential difference.

Small potential difference $\longrightarrow$ bigger potential difference

The *operational amplifier* is the basic building block for both d.c. and a.c. amplifiers. It has two inputs, known as the inverting input ($-$) and the non-inverting input ($+$), and an output. The output depends on the connections made to these inputs. Figure 3.18(*a*) shows the connections made to the operational amplifier when it is used as an inverting amplifier. The input is connected to the inverting input through a resistor R_1 and the non-inverting input is connected to ground. The output is connected to the inverting input via a resistor R_2. This is referred to a feedback since the output signal is being fed back into the input.

The operational amplifier itself has a very large transfer function, 100 000 or more, and the change in its output voltage is generally limited to about $\pm$ 10 V. Since

$$\text{transfer function} = \frac{\text{output}}{\text{input}}$$

$$\text{Input} = \frac{\text{output}}{\text{transfer function}} = \pm \frac{10}{100\,000} = \pm\, 0.0001 \text{ V}$$

The input voltage must be between + or − 0.0001 V. This is virtually zero and so point X is at virtually earth potential. For this reason it is called a *virtual earth*. The potential difference across R_1 is $(V_I - V_X)$, hence the input potential V_I can be considered to be across R_1, and so

$$V_I = I_1 R_1.$$

The operational amplifier has a very high impedance and so virtually no current flows through X into it. Hence the current through R_1 flows on through R_2. Because X is the virtual earth then, since the potential difference across R_2 is $(V_X - V_O)$, the potential difference across R_2 will be virtually $- V_O$. Hence

$$- V_O = I_1 R_2$$

Thus the transfer function which is output divided by input is given by

$$\text{transfer function} = \frac{V_O}{V_1} = - \frac{R_2}{R_1}$$

The transfer function is thus determined by the relative values of R_2 and R_1. The negative sign indicates that the output is inverted, i.e. 180° out of phase, with respect to the input. Hence the amplifier being called an inverting amplifier.

Figure 3.18(*b*) shows the operational amplifier connected as a differential amplifier. A *differential amplifier* is one which amplifies the difference between two input signals. Then

$$V_O = \frac{R_2}{R_1} (V_2 - V_1)$$

Such an amplifier finds a use in bridge circuits by amplifying the out-of-balance potential difference.

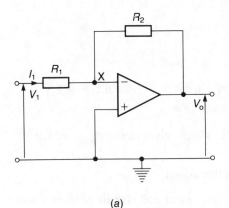

(*a*)

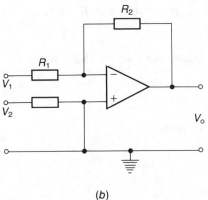

(*b*)

Fig. 3.18 (a) Inverting amplifier (b) differential amplifier

Example 5

An inverting amplifier has a resistance of 1 MΩ in the inverting input line and a feedback resistance of 10 MΩ. What is (a) the transfer function and (b) the output for an input of 0.2 V?

Answer

(a) Using the equation given above

$$\text{transfer function} = -\frac{R_2}{R_1} = -\frac{10}{1} = -10$$

(b) Since

$$\text{transfer function} = \frac{\text{output}}{\text{input}}$$

$$\text{output} = \text{transfer function} \times \text{input} = -10 \times 0.2 = -2 \text{ V}$$

Attenuation

An *attenuator* is a device which gives an output which is smaller than its input.

Large signal $\longrightarrow$ smaller signal

One way of achieving this with electrical signals is by the use of a voltage divider circuit, Fig. 3.19 shows such a circuit. The input voltage V_I is applied across the two resistors R_1 and R_2. If all the current I through R_1 passes through R_2 then

$$V_I = I(R_1 + R_2)$$

The output voltage V_O is taken from across the resistor R_2. Thus

$$V_O = IR_2$$

Dividing these two equations gives

$$\frac{V_O}{V_I} = \frac{R_2}{R_1 + R_2}$$

Example 6

What values of resistors in the Fig. 3.19 circuit will be suitable for reducing a 20 V signal to 0.5 V?

Answer

Using the equation developed above

$$\frac{V_O}{V_I} = \frac{R_2}{R_1 + R_2}$$

Hence

$$0.5(R_1 + R_2) = 20 \, R_2$$

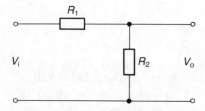

Fig. 3.19 A voltage divider circuit

$0.5 \, R_1 = 19.5 \, R_2$

If R_2 is taken to be 10 kΩ then R_1 is 390 kΩ.

The term *filtering* is used to describe the process of removing a certain band of frequencies from a signal and permitting others to be transmitted. The range of frequencies passed by a filter is known as the *pass band*, the range not passed as the *stop band* and the boundary between stopping and passing as the *cut-off frequency*. Filters are classified according to the frequency ranges they transmit or reject. A *low-pass filter* (Fig. 3.20a) has a pass band in the low frequency region, a *high-pass filter* (Fig. 3.20b) a pass band in the high-frequency region. A *band-pass* (Fig. 3.20c) filter allows a particular frequency band to be transmitted, a *band-stop* (Fig. 3.20d) filter stops a particular band.

Filters are often used with measurement systems to remove unwanted signals, e.g. mains interference. Thus the input would be the measurement signal plus unwanted signals and the output just the measurement signal.

measurement plus unwanted signals ⟶ measurement signal

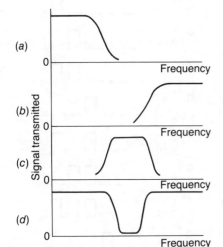

Fig. 3.20 The frequency characteristics of (a) a low-pass filter, (b) a high-pass filter, (c) a band-pass filter, (d) a band-stop filter.

Analogue to digital conversion

The output from most sensing elements tends to be in analogue form, i.e. the size of the output from the sensor is related to the size of the input. Where a microprocessor is used as part of the measurement system the analogue output from the sensor has to be converted into digital form before it can be used as an input to the microprocessor.

Analogue input ⟶ digital output

With digital signals, information is transmitted as a sequence of pulses, as illustrated in Fig. 3.21. A pulse can be considered to represent an on condition or high signal, the absence of a pulse an off or low signal. The off or low signal is represented by the binary number 0 and the on or high signal by the binary number 1 (see Table 3.1). Each such number is called a *bit*. A set of such binary numbers is called a *word*. For example, a word might be 010011. Such a word contains 6 bits.

To illustrate the action of an analogue to digital converter consider one where an input voltage of 0.1 V is required to generate 1 bit. The relationship between the input analogue voltage and the output in-bits will be as shown in Fig. 3.22.

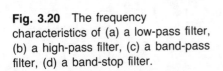

Fig. 3.21 A digital signal

Table 3.1 Binary numbers

Binary number	Signal
0	Low or off
1	High or on

Input in V	Word				Signal
0.0	0	0	0	0	
0.1	0	0	0	1	
0.2	0	0	1	0	
0.3	0	0	1	1	
0.4	0	1	0	0	
0.5	0	1	0	1	
0.6	0	1	1	0	
0.7	0	1	1	1	
0.8	1	0	0	0	
0.9	1	0	0	1	
1.0	1	0	1	0	
1.1	1	0	1	1	
1.2	1	1	0	0	
1.3	1	1	0	1	
1.4	1	1	1	0	
1.5	1	1	1	1	

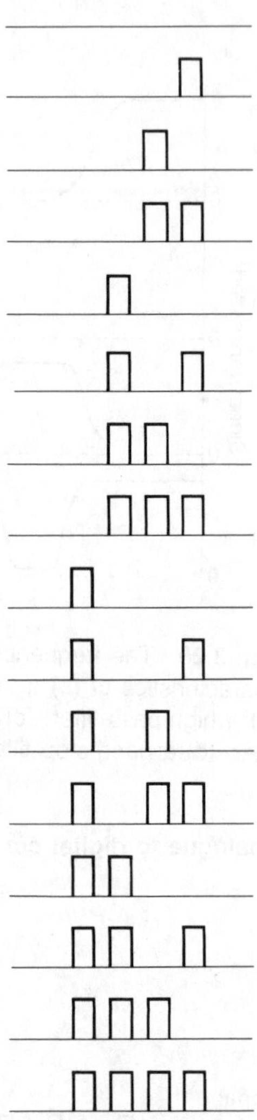

Fig. 3.22 Analogue to digital converter

With no input all the bits in the word will be 0. When the input voltage becomes 0.1 V then the first bit becomes 1. When the input voltage rises to 0.2 V, i.e. we add 0.1 V to our earlier 0.1 V input, the first bit becomes 0 and the second bit 1. What we have is

$$0001 + 0001 = 0010$$

When the input voltage rises to 0.3 V, i.e. we add another 0.1 V to our input, the first bit becomes 1 and the second bit remains at 1.

$$0001 + 0001 + 0001 = 0011$$

When the input voltage rises to 0.4 V, i.e. we add another 0.1 V to our input, the first bit becomes 0, the second bit 0 and third bit 1.

$$0001 + 0001 + 0001 + 0001 = 0100$$

The basic rules for adding binary numbers are thus

$$0 + 0 = 0$$
$$0 + 1 = 1$$
$$1 + 1 = 10$$

With the four-bit converter illustrated in Fig. 3.22 the smallest change in input that will produce a digital output is 0.1 V. This is what is termed the *resolution* of the converter. Changes in input of less than 0.1 V produce no change in digital output. For a given range of input signal the bigger the word length of the converter the better its resolution. A 4-bit word length means the signal range is broken down into 16 elements, a 6-bit word into 64 elements and an 8-bit word into 256 elements. Thus if the maximum analogue input is 10 V then with a 4-bit word converter the smallest signal that can be resolved is 10/16 = 0.625 V. With a 6-bit word this becomes 10/64 = 0.156 V and with an 8-bit word 10/256 = 0.039 V. Analogue to digital converters typically have word lengths of 8, 10, 12, 14 and 16 bits.

Example 7

An analogue to digital converter has a word length of 6 bits. What will be the output for an analogue voltage input of 150 mV if 40 mV is required to generate one bit?

Answer

We can build up the digital signal in a similar way to that used in Fig. 3.22.

Analogue input in mV	Digital output
0	000000
40	000001
80	000010
120	000011
160	000100

The input of 150 mV is insufficient to generate the 000100 but enough for 000011. Hence the output is 000011. The resolution is 40 mV.

Signal transmission

Signal transmission can take place in a number of ways. It can be, for instance, by means of mechanical, electrical, hydraulic, or pneumatic signals. *Mechanical transmission* can be by means of linked rods, levers, gearwheels, cords passing over pulleywheels, etc. Signals are transmitted as displacements. This method of transmitting signals has the disadvantage that if large distances are involved quite significant forces are needed.

An example of the mechanical transmission of signals is the Bourdon pressure gauge illustrated in Fig. 3.16. The displacement of the end of the Bourdon tube is transmitted through a linked rod to one end of what is effectively a large-diameter gearwheel. Displacement of this wheel, i.e. its rotation, is transmitted to another gearwheel and hence to a pointer which moves across a scale.

Electrical transmission is by means of current and potential difference changes through cables. It offers many advantages in that cables are flexible, easy to install and fairly cheap. They can be used for transmission over large distances, time delays between transmitting and receiving electrical signals being negligible.

An example of the electrical transmission of signals could be that involved in the monitoring of the temperature of some process by means of a thermocouple. The output from the thermocouple is a small voltage. This might be amplified to give a bigger voltage before transmission through cables to a meter in the process control room.

Hydraulic transmission is by means of pressure signals transmitted through a liquid. Hydraulic transmission has the advantages that large forces can be easily produced at the reception end, components are more rugged and resistant to shock and vibration than electrical ones, but have the disadvantages of requiring a pressurised hydraulic fluid with supply and return lines, the danger of oil leaks, and expense.

An example of the hydraulic transmission of signals could be that involved in the monitoring of the temperature of a process by a liquid-in-metal thermometer (see Fig. 2.24). When the temperature changes the liquid in the bulb of the thermometer expands and a pressure signal is transmitted through the liquid to a Bourdon pressure gauge.

Pneumatic transmission is by means of pressure signals transmitted through air contained in tubes. Pneumatic systems have the advantages of being safe if leaks occur, rugged and easily maintained, but have the disadvantages of requiring a source of compressed air, are slower acting (because air is compressible) and because of the lower pressures involved give a smaller output power than hydraulic systems.

An example of a pneumatic system is the flapper-nozzle

system described in Fig. 2.20. Displacement of the flapper is converted into a pneumatic pressure change by the sensor. This pneumatic pressure change can then be transmitted to a pressure gauge.

Example 9

Figure 3.23 shows a level-measurement system. State the form of the signals as they are transmitted through the various parts of the system.

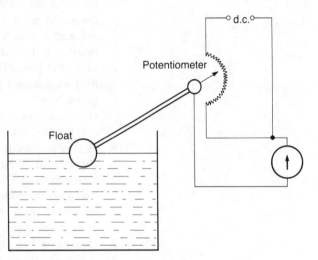

Fig. 3.23 Example 8

Answer

The input to the measurement system is the change in level of the liquid in the tank. This results in movement up or down, of the float. This then results in the up or down movement of the lever about its pivot. This then results in the movement of the potentiometer slider over its track. The potentiometer converts this motion into a changing current in an electrical circuit. This current is then displayed on a meter. Thus the earlier part of the system involves the mechanical transmission of signals, the later part electrical transmission. All the mechanical elements would be located close to the level being measured. The meter could however be located some distance away because the electrical signals can be readily transmitted over a distance.

Problems

1 Give reasons why sensing elements in a measurement system may need further processing before they can be displayed.
2 State the possible forms of input and output signals for the following signal conditioners:
   (a) Wheatstone bridge
   (b) potentiometer circuit

(c) hydraulic cylinder

3 A Wheatstone bridge, with a d.c. supply of 4.0 V, is to be used with a 100 Ω strain gauge. Initially the bridge is balanced with all the bridge arms being 100 Ω. What would be (a) the out-of-balance potential difference and (b) the change in resistance of one bridge arm to restore balance, when strain results in the strain gauge resistance increasing by 1.0 Ω?

4 A potentiometer circuit gives zero current through its galvanometer with a standard cell of e.m.f. 1.018 V when the slider is three-quarters round the track. What would be the e.m.f. which would give zero current when the slider is one-quarter of the way round the track?

5 State the possible forms of input and output signals for the following signal processors:
(a) a lever
(b) an attenuator
(c) an electronic amplifier

6 A measurement system for the measurement of pressure consists of a strain gauge pressure cell, as in Fig. 3.2, connected to a Wheatstone bridge. The output from the bridge is amplified by a differential amplifier before being displayed on a meter. State how the signal is transformed at each element.

7 Design an inverting amplifier with a voltage gain of 30.

8 An analogue signal with a maximum range of 12 V is to be converted into a digital signal using a 6-bit-word-length analogue to digital converter. What will be the change in the analogue signal that can be just resolved?

9 Give an example of a converter that could be used for each of the following signal conversions:
(a) a small displacement to a bigger displacement
(b) a small rotation to a bigger rotation
(c) a small potential difference to a bigger potential difference
(d) a resistance change to a potential difference change

10 Compare electrical, hydraulic and pneumatic methods of transmitting signals.

4 Displays

The range of display elements

There is a very wide range of elements used for the display of data from a measurement system (Fig. 4.1). They can be broadly classified into two groups: indicators and recorders. *Indicators* give an instant visual indication of the quantity being measured while *recorders* record the output signal over a period of time and give automatically a permanent record. Both indicators and recorders can be subdivided into two groups of devices, *analogue* and *digital*. An example of an analogue indicator is a meter which has a pointer moving across a scale, while a digital meter would be just a display of a series of numbers. An example of an analogue recorder is a chart recorder which has a pen moving across a moving sheet of paper, while a digital recorder has the output recorded on a sheet of paper as a sequence of numbers.

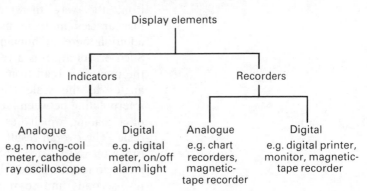

Fig. 4.1 The range of display elements

The moving-coil meter

The *moving-coil meter* is an analogue indicator which involves a pointer moving across a scale. The amount of movement of the pointer across the scale is related to the input to the meter. The basic instrument movement is a microammeter with shunts, multipliers and rectifiers being used to convert it to other ranges and measurements.

Essentially the meter movement consists of a coil in a constant magnetic field which is always at right-angles to the sides of the coil no matter what angle the coil has rotated through (Fig. 4.2). When a current passes through the coil, forces act on the coil sides and cause the coil to rotate. This rotation is opposed by springs. The result is that equilibrium is reached and a steady deflection obtained when the turning moment due to the current through the coil is exactly balanced by that produced by the springs. The result of this is that the angular movement of the coil is proportional to the current.

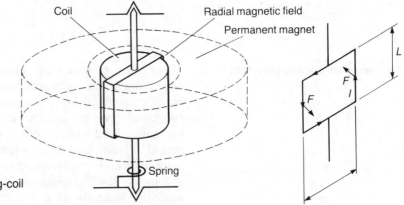

Fig. 4.2 The basis of the moving-coil meter

Moving-coil meters generally have resistances of the order of a hundred ohms. The accuracy of such a meter depends on a number of factors. Among them are temperature, the presence nearby of magnetic fields or magnetic materials like iron, the way the meter is mounted, bearing friction, inaccuracies in scale marking during manufacture, etc. In addition there are human errors involved in reading the meter. Such errors arise as a result of parallax when the position of the pointer is read from an angle other than directly at right-angles to the scale (see Ch. 1) and also from errors in interpolating between scale markings, i.e. estimating the value of a reading which lies between two marked points on the scale. The overall result is that accuracies are generally of the order of ± 0.1% to ± 5%. The time taken for a moving-coil meter to reach a steady deflection is typically in the region of a few seconds and so it cannot respond to rapidly changing signals.

The digital meter

The *digital meter* gives its reading in the form of a sequence of digits. Such a form of display eliminates parallax and interpolation errors and can give accuracies as high as ± 0.005%. Typically such a meter has a resistance of the order of 10 MΩ. The digital meter is based on the digital voltmeter, of which there are a number of forms. All the forms can however be considered to be a form of analogue to digital converter connected to a counter (Fig. 4.3). At some instant of time the analogue voltage is sampled and the value at that time converted into a digital signal and displayed. Some forms of digital voltmeter can take samples every one-thousandth of a second while others are much slower. The shorter the time interval between samples the more useful the meter is for rapidly changing signals.

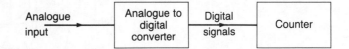

Fig. 4.3 Digital voltmeter principle

Example 1

A digital voltmeter specification includes the statement: sample rate approximately 5 readings per second. What is the significance of this?

Answer

The sample rate of 5 readings per second means that every 0.2 s the input voltage is sampled. It is the time taken for the instrument to process the signal and give a reading. Thus if the input voltage is changing at a rate which results in significant changes during 0.2 s then the voltmeter reading can be in error.

On-off displays

There are many situations where all that is required of a measurement system display is that it should indicate when some particular condition is reached. Thus there might be a display which gives a signal when the temperature reaches a particular value or falls to some other value. These may be based on the use of a resistance element or a thermocouple and the display might be a red light that comes on when the temperature value is reached.

Alarm indicators take an analogue input from some transducer (possibly via a signal conditioner) and turn it into an on-off signal for some display, e.g. a bell, a horn, a klaxon, a coloured light, a flashing light, a backlighted display (the light comes on behind a message on a screen).

Analogue chart recorders

With analogue chart recorders the input signal results in a marking mechanism marking a point on a chart. By either moving the marking mechanism or the chart in a controlled way with time an analogue record can be obtained of how the input signal varied with time.

The *direct-reading type* of chart recorder involves the

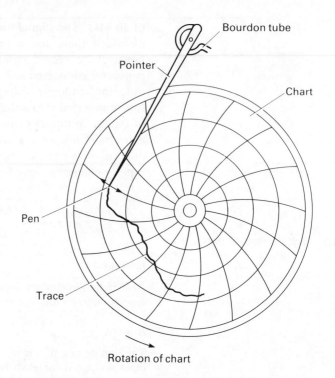

Bourdon tube

Pointer

Chart

Pen

Trace

Rotation of chart

Fig. 4.4 Direct-reading chart recorder

marking mechanism, a pointer with a pen at its end, being directly moved by the measurement system (Fig. 4.4). Thus in the case of a pressure recorder the displacement of the end of a Bourdon tube may be used to move the pointer directly. In the case of a temperature recorder a bimetallic strip may be used to move the pointer directly. This type of instrument has a circular chart which rotates at a constant rate. The trace produced on the chart is thus a graph of pressure or temperature against time with the pressure or temperature being measured radially outwards from the chart centre and time circumferentially. The radial lines are curved because the movement of the pen at the end of the pointer is in the arc of a circle. Because of the curved nature of the radial and circumferential graph axes there is difficulty in estimating values which lie between marked lines on the chart. There is also particular difficulty in determining values for traces close to the centre of the chart where the radial lines are very close together. Such a chart also has the problem that after one revolution traces on the chart become superimposed, hence such a chart has to be replaced before this occurs if confusion is to be avoided.

A direct-reading chart recorder may have more than one pointer and pen so that it can record simultaneously more than one measurement. The accuracy is generally of the order of ± 0.5% of the full-scale deflection of the signal. The chart

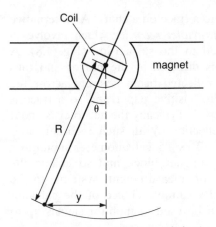

Fig. 4.5 Galvanometric type of chart recorder

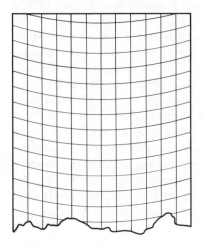

Fig. 4.6 Curvilinear chart paper

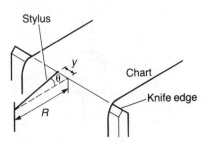

Fig. 4.7 Knife-edge recorder

drive generally gives one complete rotation in 12 h, 24 h or 7 days.

The *galvanometric type* of chart recorder (Fig. 4.5) works on the same principle as the moving-coil meter movement described earlier in this chapter. A pointer with a pen at its end is used to produce an ink trace on the chart.

If R is the length of the pointer and θ the angular deflection of the coil, then the displacement y of the pen is

$$y = R \sin \theta$$

Since the angle θ is proportional to the current through the coil, then the relationship between the current and the displacement is non-linear. For a linear relationship we would have required $y = R\theta$. However, if the angular deflections are restricted to less than $\pm 10°$ then the error that occurs by assuming the linear relationship is quite small. The error is the differences in values between θ in radians and $\sin \theta$. With such a restriction the error due to this non-linearity is less than 0.5%. A greater problem is however the fact that the pen moves in an arc rather than a straight line. Thus curvilinear paper (Fig. 4.6) is used for the plotting. However there are difficulties in estimating the values of points between the curved lines.

An alternative form of 'pen' chart recorder which leads to rectilinear charts rather than curvilinear charts is the *knife-edge recorder* (Fig. 4.7). One version of this uses heat-sensitive paper which moves over a knife edge and a heated stylus instead of an ink pen. The paper may be impregnated with a chemical that shows a marked colour change when heated by contact with the stylus or the stylus burns away temperature-sensitive outer layers which coat the paper. The use of the knife edge avoids the curved trace but the non-linear relationship between θ and displacement y still exists. The length of trace y produced on the paper by a deflection θ is

$$y = R \tan \theta$$

The non-linearity error is slightly greater than for the pen form of recorder. If deflections are restricted to less than $\pm 10°$ the error is less than 1%.

Usually with the pen or knife-edge form of galvanometric recorder there is some electronic amplification of the input signal. This leads to sensitivities which are generally of the order of 1 centimetre of pen displacement per millivolt input, resistances of about 10 kΩ, accuracies of the order of $\pm 2\%$ of the full-scale deflection and a speed of response which enables the recorder to be used for d.c. signals and a.c. signals up to about 50 Hz.

There are a number of ways by which the movement of the

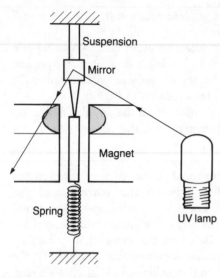

Fig. 4.8 Ultraviolet galvanometric recorder

coil may be transformed into a trace on a chart. An alternative to the pen form is the *ultraviolet recorder* which involves a small mirror being attached to the suspension (Fig. 4.8). A beam of ultraviolet light is directed at the mirror and thus when the coil rotates the reflected beam is swept across the chart. The chart uses photosensitive paper and so a trace is produced when it is developed. Typically the accuracy is about ± 2% of the full-scale deflection. With such recorders it is quite common to have 6, 12 or 25 galvanometer mountings side by side in the same magnet block and so enable the outputs from a number of measurement systems to be simultaneously displayed. By suitable choice of the galvanometer coil the recorder can be used to monitor signals up to about 13 kHz.

Figure 4.9 illustrates the general principles of the potentiometric recorder. Such a recorder is sometimes referred to as a *closed-loop recorder* or a *closed-loop servo recorder*. The position of the pen is monitored by means of a slider which moves along a linear potentiometer. The position of the slider determines the potential applied to an operational amplifier. The amplifier subtracts the pen signal from the signal from the measurement system. The output from the amplifier is thus a signal related to the difference between the pen and measurement system signals. This signal is used to operate a motor which in turn controls the movement of the pen across the chart. The result of all this is that the pen ends up moving to a position where the result is no difference between the pen and measurement system signals. In other words, the position of the pen is adjusted until it gives a signal which has the same

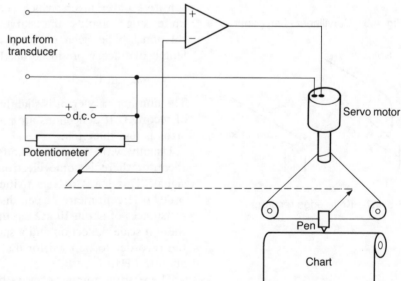

Fig. 4.9 Potentiometric recorder

value as that from the measurement system. The position of the pen is then a measure of the measurement system signal.

Potentiometric records typically have high input resistances, higher accuracies (about $\pm 0.1\%$ of full-scale reading) than galvanometric recorders but considerably slower response times. They can thus only really be used for d.c. or slowly changing signals. Because of friction there is a minimum current required to get the motor operating. There is thus some error due to the recorder not responding to a small sensor signal. This error is known as the *dead band*. Typically it is about $\pm 0.3\%$ of the range of the instrument. Thus for a range of 5 mV the dead band error amounts to ± 0.015 mV.

Example 2

A chart recorder is to be used to monitor the temperatures of liquids in a number of vessels. The temperatures do not vary rapidly with time. A potentiometric chart recorder has been suggested. Would this be suitable?

Answer

Such a recorder can only be used with signals which do not vary rapidly with time and so would be suitable.

Example 3

A chart recorder has a chart width of 150 mm and is being used at a chart speed of 120 mm per hour. A full-scale deflection is produced by 50 mV. What will be (a) the input when there is a deflection of 120 mm, and (b) the time between two events if the distance between them on the chart is 500 mm?

Answer

(a) A deflection of 150 mm is produced by an input of 50 mV. Hence a deflection of 1 mm requires an input of 50/150 mV and so a deflection of 120 mm has an input of

$$\text{input} = \frac{120 \times 50}{150} = 40 \text{ mV}$$

(b) A chart speed of 120 mm per hour means a time between the two events of

$$\text{time} = \frac{500}{120} = 4.2 \text{ h}$$

Cathode ray oscilloscope

Figure 4.10 shows the basic features of the *cathode ray tube*. The tube consists of an electron gun to produce a focused beam of electrons on the screen and a deflection system for moving the beam in the vertical and horizontal directions. The electron gun has the following elements:

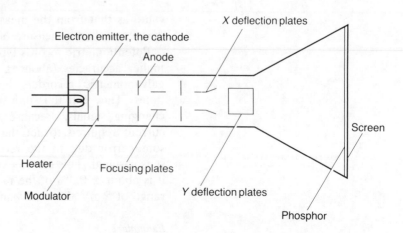

Fig.4.10 Cathode ray tube

1 *Electron production* Electrons are produced by a current being passed through a filament to heat the cathode.
2 *Brilliance control* The number of the electrons which are to form the electron beam, and so determine the brilliance of the spot on the fluorescent screen, is determined by a potential applied to an electrode immediately in front of the cathode. This electrode is called the modulator.
3 *Electron acceleration* The electrons are accelerated down the tube by a potential difference between the cathode and the anode.
4 *Focusing* An electron lens is used to focus the beam so that when it reaches the phosphor-coated screen it forms a small luminous spot. The focus is adjusted by changing the potential of electrodes relative to that of the earlier electrodes.

The deflection system consists of two elements:

1 *Y deflection* The beam can be deflected in the vertical or *Y* direction by a potential difference applied between the *Y* deflection plates.
2 *X deflection* A potential difference between the horizontal or *X* deflection plates will cause the beam to deflect in the *X* direction.

The oscilloscope is an example of an instrument employing a cathode ray tube, another being the TV monitor. With the oscilloscope the screen has a rectangular grid scale to enable measurements to be made of the deflection of the luminous spot.

A signal applied to the *Y deflection plates* causes the electron beam, and hence the spot on the screen, to move up or down in the vertical direction. A range selector switch enables different deflection sensitivities to be obtained. A general-purpose oscilloscope is likely to have sensitivities which vary between 5 mV per scale division to 20 V per scale

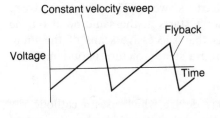

Fig. 4.11 Sawtooth waveform

division. The input generally has an a.c. or d.c. mode of operation switch. The a.c. mode introduces a blocking capacitor in the input line to eliminate d.c. components of the signal. In the a.c. mode typically signals with frequencies from about 2 Hz–10 MHz can be displayed, when in its d.c. mode from d.c. to 10 MHz.

The *X deflection plates* are generally used with an internally generated signal, a voltage with a sawtooth waveform (Fig. 4.11). This sweeps the luminous spot on the screen from left to right at a constant velocity with a very rapid return ,i.e. flyback. This return is too fast to leave a trace on the screen. The constant velocity movement from left to right means that the distance moved in the X direction is proportional to the time elapsed. Hence the sawtooth waveform gives a horizontal time axis, i.e. a *time base*. Typically an oscilloscope will have range of time bases from about 1 s per scale division to 0.2 μs per scale division.

For an input signal to give rise to a steady trace on the screen when there is a signal which is varying with time in a regular manner, it is necessary to synchronise the timebase and the input signal so that every time the trace starts at the left-hand side of the screen it starts at the same point on the input signal. For this purpose a *trigger circuit* is used. The trigger circuit can be adjusted so that it responds to a particular voltage level and also whether the voltage is increasing or decreasing. This means that for a periodic signal input the trigger circuit can be set to respond to particular points in its cycle (Fig. 4.12). Whenever the input reaches the set point a

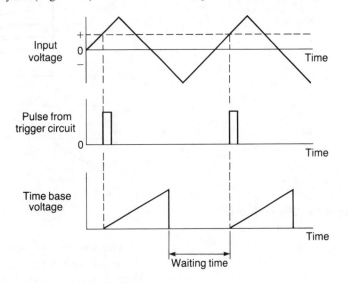

Fig. 4.12 Triggering

pulse is produced by the trigger circuit. This pulse triggers the time base into action. Until such a pulse is received the time base waits and does not start its sweep. The time base sweep across the screen thus always starts at the same point on the input signal. The result is that successive scans of the input signal are superimposed and a steady picture is obtained.

Monitors

A monitor is just a display device which uses a cathode ray tube, to display letters of the alphabet and numbers (so called alphanumeric data), graphic and pictorial data. Sawtooth signals are applied to both the X and the Y deflection plates (Fig. 4.13). The Y input moves the spot relatively slowly from the top to the bottom of the screen before a rapid flyback to the top of the screen. During the time taken for the Y deflection to go from top to bottom of the screen the X deflection input has moved the spot from left to right across the screen many times. The result of both these inputs is that the spot zig-zags down the screen. During its travel the electron beam is switched on or off by an input to the modulation electrode. The result is that a 'picture' can be painted on the screen.

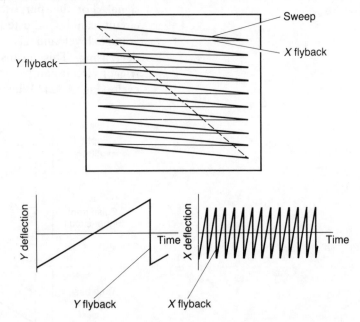

Fig. 4.13 Raster display

The above is a description of the basis of a monochrome monitor with what is called a *raster display*. A colour monitor has a screen coated with dots of three different types of phosphor. One type emits red light, one green light and the other blue light. Three electron guns are used, one for each type of phosphor. Dots of the these three types are arranged in clusters. If the three beams energise all three the appearance is of white light, if just the red phosphor is energised, red light.

Magnetic-tape recorders

The magnetic-tape recorder can be used to record both analogue and digital signals. It consists of:

- a recording head which responds to the input signal and produces corresponding magnetic patterns on magnetic tape
- magnetic tape which retains the magnetic pattern produced by the recording head
- a replay head to do the converse job and convert the magnetic patterns on the tape to electrical signals
- a tape transport system which moves the magnetic tape at a constant speed under the record and replay heads
- signal conditioning elements such as amplifiers and filters

The recording head consists of a coil wound on a core of ferromagnetic material (Fig. 4.14). The core is in the form of a loop which is almost closed, there being a small gap between the two ends of the core. The magnetic tape moves so that it effectively bridges this gap. When electrical signals are fed to the coil round the core it becomes magnetised. So also does that part of the magnetic tape in the region of the gap. The magnetic tape is a flexible plastic base coated with a ferromagnetic powder. Hence a magnetic record is produced of the electrical input signal.

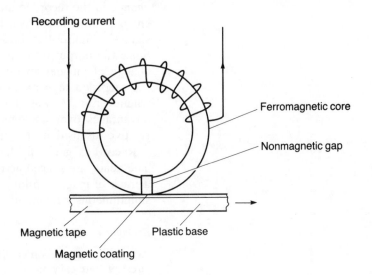

Fig. 4.14 Basis of a magnetic recording head

The recording head and the replay heads have similar forms of construction. When a piece of magnetised tape bridges the gap of the replay head then the effect is just like bringing a permanent magnet near a piece of unmagnetised iron. It also becomes magnetised. The core however passes through a coil of wire. These changes in the core induce e.m.f.s in the coil. Thus the output from the coil is an electrical signal which is related to the magnetic record on the tape.

The above is a discussion of what is termed *direct-recording*. With this form the input signal directly determines the magnetism recorded on the tape. Thus if the input was a constant d.c. signal then the tape would have a constant level of magnetism. However the replay head cannot respond to a constant level of magnetism, only changes in magnetism. This is because the replay head coil has an e.m.f. induced in it by electromagnetic induction and an e.m.f. is only produced when the flux linking the coil changes. Thus this form of recording and playback cannot handle d.c. signals.

An alternative to this is *frequency modulation* recording. With this the input signal is used to vary the frequency of an alternating current, the amount by which the frequency is changed being related to the size of the input signal. Thus the signal which is recorded is always an alternating signal. The replay head extracts the modulated signal from the tape and it is then converted back into an unmodulated signal. Frequency modulation recording can thus be used with d.c. signals.

Magnetic tape recorders can be used for recording analogue or digital signals. A commonly used digital recording method is the *non-return-to-zero* (*NRZ*) method. With this system no change in the recorded magnetism is used to represent 0 and a change in the magnetism 1 (see Ch. 3 for a discussion of digital signals). Figure 4.15 illustrates this for the number 0110101. Since the output from the replay head depends on the rate of change of magnetism on the tape, outputs only occur where the recorded tape has a change. Thus the output is a pulse whenever a 1 was recorded. Digital recording has the advantages over analogue recording of higher accuracy and relative insensitivity to tape speed.

Recorders generally have more than one recording head. The heads are spaced across the tape and thus each lays down a track of magnetisation. Thus several different signals can be simultaneously recorded.

Example 4

A magnetic-tape recorder offers the option of being operated in the direct or frequency modulated modes. Which mode should be used if the signal to be recorded is slow varying d.c.?

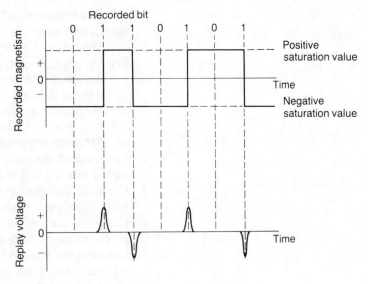

Fig. 4.15 Non-return-to-zero recording

Answer

The direct mode of operation is not suitable for slow varying d.c. The frequency modulated mode can operate down to d.c. and so is the choice.

Digital printers

Analogue chart recorders give records in the form of a continuous trace, digital printers give records in the form of numbers, letters or special characters. Such printers are known as *alphanumeric printers*. There are a numbers of versions of such printer, the most commonly used one being the *dot-matrix printer*. With such a printer the print head consists of either nine or twenty-four pins in a vertical line. Each pin is controlled by an electromagnet which when turned on propels the pin onto an inking ribbon. This impact forces a small blob of ink onto the paper behind the ribbon. A character is formed by moving the print head across the paper and firing the appropriate pins. The 24-pin head uses more dots to form the character and so gives a better-quality image, the individual dots being not discernible to the naked eye. With the 9-pin head this is not the case.

Problems

1 The following are extracts from the specifications of instruments. Explain the significance for the use of such instruments of the terms and values quoted.
   (a) Digital meter
       Sample rate 5 readings/second
   (b) Closed-loop servo recorder

Dead band ± 0.2% of span
(c) Cathode ray oscilloscope
Time base 1 s/div to 0.2 μs/div
(d) Moving-coil meter
Accuracy ± 1% of full-scale deflection
Ranges 0–5 mA, 0–50 mA, 0–500 mA
(e) Magnetic-tape recorder
Direct record/reproduce
FM record/reproduce

2 List some of the main sources of inaccuracy that can occur with a moving-coil meter.

3 Compare the characteristics of the (a) galvanometric (b) knife-edge and (c) ultraviolet forms of recorders.

4 Compare, for ease of reading, the forms of charts that can be produced by different forms of chart recorder.

5 Describe the basic principles involved in the operation of:
(a) the cathode ray tube
(b) the magnetic-tape recorder
(c) a galvanometric recorder

6 A chart recorder has a chart speed of 20 mm per hour. How far apart will be two events that took place 15 minutes apart?

7 A chart recorder has a chart width of 120 mm and gives a full-scale deflection with 10 mV. What will be the deflection with 2 mV?

5 Pressure measurement

Pressure is defined as being the force per unit area, i.e.

$$\text{pressure } P = \frac{\text{force } F}{\text{area } A}$$

The unit of pressure is the pascal (Pa), with 1 Pa being when a force of 1 N acts over an area of 1 m^2.

The term *absolute pressure* is used for the pressure measured relative to zero pressure, the term *gauge pressure* for the pressure measured relative to atmospheric pressure. The gauge pressure thus states how much more than the atmospheric pressure a pressure is. At the surface of the earth the atmospheric pressure is generally about 100 kPa. This is sometimes referred to as a pressure of 1 bar.

Absolute pressure = gauge pressure + atmospheric pressure

Example 1

A pressure-measurement system gives a gauge pressure of 5 kPa. What is the absolute pressure if the atmospheric pressure is 100 kPa?

Answer

Using the equation given above

Absolute pressure = gauge pressure + atmospheric pressure
= 100 + 5
= 105 kPa

Pressure in a fluid

Two fundamental laws of fluid pressure are:

- The pressure at a point in a fluid is the same in all directions.
- The pressure exerted on any surface in a fluid is always at right-angles to the surface.

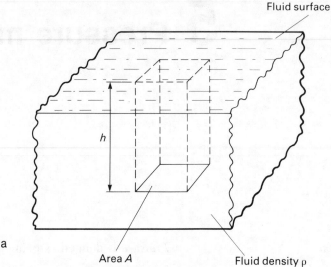

Fig. 5.1 Pressure due to weight of a fluid

Consider the pressure at some depth in a fluid at rest (Fig. 5.1) due to just the weight of fluid above it, i.e. the gauge pressure. If we consider a surface of area A at that depth then the force acting on that area is the weight of the fluid directly above it. The volume of a column of liquid of cross-sectional area A and height h is hA. If the fluid has a density ρ then the weight is $hA\rho g$, where g is the acceleration due to gravity. Thus the pressure is

$$\text{pressure } P = \frac{\text{weight of fluid}}{\text{area}}$$

$$= \frac{hA\rho g}{A}$$

Hence

$$P = h\rho g$$

Example 2

What is the gauge pressure in a liquid of density 1000 kg/m³ at depths of (a) 0.020 m, (b) 0.100 m? Take the acceleration due to gravity to be 9.81 m/s².

Answer

(a) Using the equation given above

$$P = h\rho g$$

$$= 0.020 \times 1000 \times 9.81 = 196 \text{ Pa}$$

(b) Using the same equation,

$$P = 0.100 \times 1000 \times 9.81 = 981 \text{ Pa}$$

Manometers

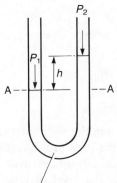

Density ρ

Fig. 5.2 U-tube manometer

Figure 5.2 shows the simple *U-tube manometer*. It consists of a U-tube containing a liquid. When the pressures in the gases above the liquid in the two limbs are the same, i.e. $P_1 = P_2$, then the height of the liquid in each limb of the U is the same. When there is a pressure difference between the gases above the liquid in the two limbs a difference in heights of the liquid occurs. This difference in height h is a measure of the difference in pressure.

The pressure at height AA in the two limbs is the same because for a stationary liquid there cannot be a difference in pressure between two points at the same horizontal level. If there was a difference then the liquid would flow between the two points until there was no difference. For the left hand limb this pressure is P_1, for the right-hand limb it is $P_2 + h\varrho g$. Thus

$$P_1 = P_2 + h\varrho g$$

where ϱ is the density of the manometric liquid and g the acceleration due to gravity. Hence

$$\text{Pressure difference} = P_1 - P_2 = h\varrho g$$

If one of the limbs is open to the atmosphere then the pressure difference is that between the gas pressure and the atmosphere and is the gauge pressure.

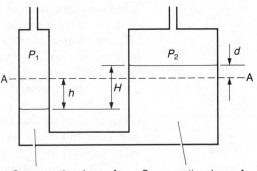

Fig. 5.3 An industrial manometer

Cross-sectional area A_1 Cross-sectional area A_2

An industrial form of manometer is shown in Fig. 5.3. The arrangement is still a form of U-tube but one of the limbs has a much greater cross-sectional area than the other. When the pressures P_1 and P_2 are equal then the level of the liquid is AA. A difference in pressure between the two limbs causes a difference in liquid level. For this difference in level to occur liquid has to flow from one limb to the other. For such an arrangement

$$\text{pressure difference} = P_1 - P_2 = H\varrho g$$

But $H = h + d$, where h and d are the changes in level in each limb from the level AA that existed when there was no pressure difference. Thus

$$\text{pressure difference} = (h + d)\varrho g$$

The volume of liquid leaving one limb must equal the volume entering the other. Hence

$$A_1h = A_2d$$

where A_1 and A_2 are the cross-sectional areas of the two limbs. Hence

$$\begin{aligned}\text{pressure difference} &= [(A_2d/A_1) + d]\varrho g \\ &= [(A_2/A_1) + 1]d\varrho g \\ &= [\text{a constant}]d\varrho g\end{aligned}$$

Thus the movement of the liquid level d in the wide tube from its initial zero level is proportional to the pressure difference. This form of manometer has thus an advantage over the simple U-tube manometer in that only the level of liquid in one limb has to be measured and this measurement is always made from the same fixed point, the level AA. A scale could thus be fixed to the tube with its zero at level AA. It is, however, more usual to determine the displacement by using a float and lever system to move a pointer across a scale (see Ch. 6 for details of level measurement).

Manometers, with an appropriate manometric liquid, can be used to measure pressure differences of the order of 20 Pa to 140 kPa. Water, alcohol or mercury are commonly used as the manometric liquid. The measurements involve the vertical height difference between liquid levels and so errors can occur if the height measured is not truly vertical. The density of the manometric liquid is required and account may need to be taken of the fact that the density depends on the temperature (see later this chapter). Thus errors will occur if the density value at, say, 18°C is used when the temperature is perhaps 22°C. The acceleration due to gravity is also required and errors will occur if the correct value for where the measurement is being made is not used. The acceleration depends on the geographical latitude (see Table 5.1 for some typical values) and the height above the earth's surface. The acceleration due to gravity decreases by about 3.1×10^{-6} m/s^2 for each metre above sea level. There are also difficulties in obtaining an accurate reading of the level of a liquid in a tube due to the meniscus of the manometric liquid. Taking these factors into account the accuracy of pressure difference measurement using a manometer is typically about $\pm 1\%$.

Table 5.1 Acceleration due to gravity values

Place	Latitude	g in m/s^2
Equator	0°	9.7803
Madras, India	13° 4′ N	9.7828
Hong Kong	22° 18′ N	9.7877
Cape Town, S. Africa	33° 56′ S	9.7966
Melbourne, Australia	37° 50′ S	9.7999
New York, USA	40° 48′ N	9.8027
London, England	51° 28′ N	9.8120
Copenhagen, Denmark	55° 41′ N	9.8156
Pole	90°	9.8322

Example 3

What is the pressure difference indicated by a U-tube manometer if there is a difference in liquid level between the two limbs of 10 cm? The liquid has a density of 1000 kg/m^3 and the acceleration due to gravity is 9.8 m/s^2.

Answer

Using the equation developed above

$$\text{pressure difference} = h\varrho g$$
$$= 0.010 \times 1000 \times 9.8 = 98 \text{ Pa}$$

Temperature correction for density change

The correction that has to be made for the effect of temperature on the density of the manometer liquid is derived as follows. If we consider a mass m of the liquid at 0°C then it has a volume V_0 and a density ϱ_0 related by

$$\text{density} = \text{mass/volume}$$

and so

$$\text{mass} = \text{density} \times \text{volume}$$
$$m = \varrho_0 V_0$$

At temperature θ the same mass of liquid will have a volume V_θ and density ϱ_θ.

$$m = \varrho_\theta V_\theta$$

Hence

$$\varrho_\theta V_\theta = \varrho_0 V_0$$

But the volume at temperature θ is related to the volume at 0°C by

$$V_\theta = V_0(1 + \gamma\theta)$$

where γ is the coefficient of cubical expansion of the liquid. Hence

$$\varrho_\theta = \frac{\varrho_0 V_0}{V_\theta}$$

$$\varrho_\theta = \frac{\varrho_0}{1 + \gamma\theta}$$

Thus neglecting any other corrections, the pressure when measured by a manometer at temperature θ when the manometer liquid density at 0°C is known is

$$P = H\varrho_\theta g = \frac{H\varrho_0 g}{1 + \gamma\theta}$$

Example 4

The density of mercury at 0°C is 13 595 kg/m³. What will be its density at 20°C if the coefficient of cubical expansion is 0.000 181/°C?

Answer

Using the equation developed above

$$\varrho_\theta = \frac{\varrho_0}{1 + \gamma\theta}$$

$$= \frac{13\ 595}{1 + 0.000\ 181 \times 20}$$

$$= 13\ 546 \text{ kg/m}^3$$

An alternative method of arriving at the density of mercury is to use tables which give density values at a range of temperatures.

Bourdon tubes

The Bourdon tube (see Ch. 2) is a very widely used sensor for the measurement of pressure. The Bourdon tube may be in the form of a C, a flat spiral, a helical spiral or twisted. Whatever the form, an increase in the pressure inside the tube results in the tube straightening out to an extent which depends on the pressure change. The displacement of the end of the tube may be used directly to move a pointer across a scale (see Fig. 3.16) or to move the slider of a potentiometer (Fig. 3.4) or to move the core iron rod in a linear variable differential transformer (Fig. 5.4). These last two will give an electrical output related to the pressure.

Typically, Bourdon tube measurement systems are used for pressure differences in the range 10 kPa to 100 MPa. The range depends on the form of the Bourdon tube and the material from which it is made. C-shaped tubes with an oval, almost rectangular, cross-section and made from brass or

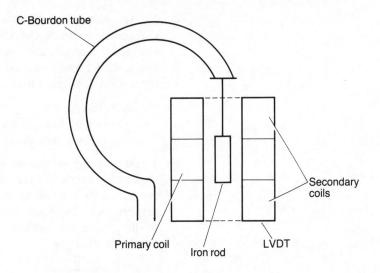

Fig. 5.4   Bourdon tube with a linear variable differential transformer

phosphor bronze have a pressure range from about 35 kPa to 100 MPa. Spiral and helical tubes have greater sensitivity and resolution than the simple C shape. However this does have the effect of reducing the maximum pressure that can be measured to about 50 MPa. It also significantly increases the cost. Bourdon tubes are robust with an accuracy of about ± 1% of full-scale reading.

Diaphragms

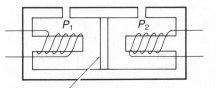

Fig. 5.5   Variable reluctance differential pressure sensor

Pressure gauges based on the use of a diaphragm (see Ch. 2) depend for their action on a difference in pressure between the two sides of the diaphragm causing it to deform, i.e., bow out to one side or the other. The greater the pressure difference the greater the amount of deformation. There are a number of methods used to detect and measure the deformation. The following are some commonly used ones.

1   *As a change in reluctance (Fig. 5.5)*   The displacement of the central part of the diaphragm gives rise to a change in reluctance, increasing that on one side of the diaphragm and decreasing it on the other. Reluctance in magnetism is rather like resistance in electricity. Increasing the resistance in an electrical circuit decreases the current, increasing the reluctance in a magnetic circuit decreases the magnetic flux in that circuit. Thus the diaphragm bowing to one side increases the magnetic flux through the coil on that side and decreases it through the coil on the other side.

Pressure change
$\longrightarrow$ diaphragm displacement change
$\longrightarrow$ change in reluctance

2 *As a change in capacitance (Fig. 2.7)* The displacement of the central part of the diaphragm relative to a fixed plate changes the capacitance between the diaphragm and the fixed plate. The capacitance of a pair of plates depends on their separation; changing that changes the capacitance.

Pressure change
⟶ diaphragm displacement change
⟶ capacitance change

3 *As a change in resistance (Fig. 3.2)* Strain gauges could be attached directly to the diaphragm, or as in Fig. 3.2 to a cantilever which is bent when the central part of the diaphragm is displaced. The result of either method is that the strain gauges are subject to a strain and consequently their resistance changes.

Pressure change
⟶ diaphragm displacement change
⟶ bending of cantilever
⟶ change in strain for strain gauges
⟶ change in resistance of strain gauges

With reluctance or capacitance changes the signal conditioner used is likely to be an a.c. bridge with its out-of-balance signal being amplified. With the strain gauges the signal conditioner is likely to be a Wheatstone bridge (see Ch. 3), the out-of-balance signal being amplified. Diaphragm instruments in general are used in the range 1 Pa to 100 MPa and have an accuracy of about ± 0.1%. Capacitive versions tend to have ranges of about 1 Pa to 200 kPa, reluctance versions 1 Pa to 100 MPa and strain gauge versions 100 kPa to 100 MPa. Capacitive, reluctance and strain gauge versions can be used with pressure fluctuations at frequencies up to about 1 kHz.

Semiconductor strain gauges can be used to detect the movement of a diaphragm. While such gauges could be cemented to the surface of the diaphragm, it is now more customary to use a silicon sheet as the diaphragm and introduce doping material into the silicon, at appropriate places, and so produce the strain gauges integral with the diaphragm (Fig. 5.6). Such gauges can be used with a

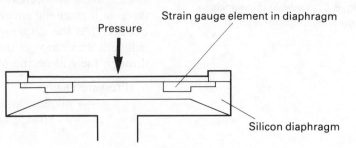

Fig. 5.6 Semiconductor strain gauge pressure cell

Wheatstone bridge. Typically such a gauge can respond to pressure differences up to about 100 kPa with an accuracy of about ± 0.5%. With the Wheatstone bridge the typical out-of-balance voltage is a few millivolts for each kilopascal pressure difference.

Capsules and bellows

A capsule can be considered to be just a form of diaphragm. Figure 5.7 shows one form of pressure gauge used for the measurement of the atmospheric pressure. It is called the *aneroid barometer*. It consists of a sealed capsule from which the air has been partially removed. This means that the air pressure inside the capsule is less than the atmospheric pressure being measured. Changes in atmospheric pressure cause the surface of the capsule to become displaced. This displacement is magnified by a system of levers and results in the movement of a pointer across a scale. See later this chapter for further discussion.

Change in atmospheric pressure
    ⟶ displacement of capsule surface
        ⟶ movement of levers
            ⟶ pulling on cord
                ⟶ movement of pointer across scale

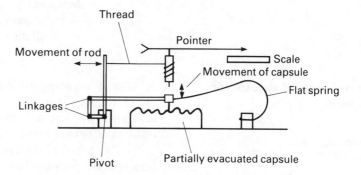

Fig. 5.7   Aneroid barometer

When the pressure inside a bellows changes the bellows will change in length. This length change is related to the pressure. Figure 5.8 shows one form of *bellows pressure sensor*. Pressure changes in the bellows cause the sealed end of the bellows to become displaced. This in turn moves the core rod in a linear variable differential transformer (see Ch. 2) and so gives an output which is related to the pressure in the bellows. Another form uses the movement of the end of the bellows to move the slider of a potentiometer. Bellows instruments are used to measure pressure differences in the range 200 Pa to 1 MPa with an accuracy of about ± 0.1%. They have poor zero stability.

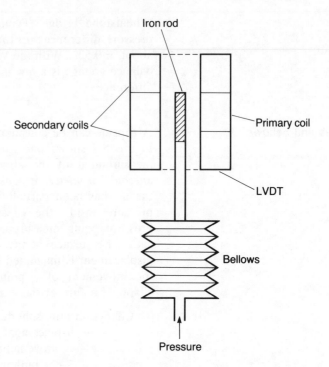

Fig. 5.8 Bellows pressure sensor

Change in pressure
⟶ displacement of end of bellows
⟶ movement of rod in LVDT
⟶ change in electrical output from LVDT

Measurement of atmospheric pressure

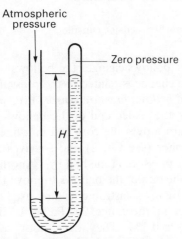

Fig. 5.9 Atmospheric pressure

A U-tube manometer with a vacuum, i.e., zero pressure, above the liquid in one limb and the atmospheric pressure acting on the liquid in the other limb would give a height difference between the manometer liquid levels of H (Fig. 5.9). The pressure difference between the two limbs is the atmospheric pressure and thus the atmospheric pressure is $H\varrho g$, where ϱ is the density of the manometer liquid and g the acceleration due to gravity. With mercury as the manometer liquid H tends to be about 760 mm. A more convenient form of the U-tube manometer for the measurement of atmospheric pressure is shown in Fig. 5.10. This is essentially the U-tube minus the glass of one of the limbs. The atmospheric pressure acting on the surface of the mercury in the open vessel is thus

atmospheric pressure = $H\varrho g$

This is the height of a column of liquid of density ϱ that can be supported by the atmospheric pressure. Instruments for the measurement of the atmospheric pressure are known as *barometers*.

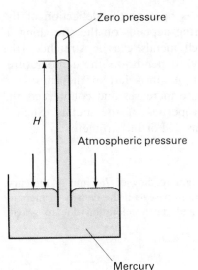

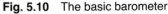

Fig. 5.10 The basic barometer

The *Fortin barometer* (Fig. 5.11) is a commercial form of the basic mercury manometer. This has the glass tube in a protective jacket and scales enabling the height of the mercury surface in the closed tube to be directly read. The vernier scale is adjusted in position so that it just coincides with the mercury surface. It can then be read in conjunction with the fixed scale to give an accurate value for the height of the mercury surface. Before such a reading is made it is essential to ensure that the zero of the fixed scale coincides with the mercury level in the lower reservoir. This mercury is in a container which has a leather bag for its lower surface. By rotating the zero adjustment screw the bag can be distorted and so the level of mercury in the reservoir raised or lowered until it coincides with the tip of the zero index. When this occurs the scale is correctly zeroed with the mercury surface. For accurate values of the atmospheric pressure a number of corrections have to be made to the reading (see later this chapter).

Another method of measuring the atmospheric pressure is the *aneroid barometer* (see Fig. 5.7 and the discussion earlier in this chapter). This consists of a partially evacuated capsule which contracts or expands as a result of changes in the amospheric pressure. If the capsule is completely evacuated corrections have to be made for the temperature at which the

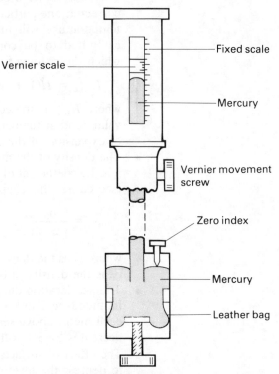

Fig. 5.11 The Fortin barometer

measurement is made. This is because the deflection of the capsule material and the spring depends on the stretching of metals. The ease with which metals can be stretched (the tensile modulus of elasticity) depends on the temperature. However if a small amount of air is left in the capsule it expands when the temperature increases and counteracts the change in the stretching properties of the metals. Such a barometer is calibrated against a Fortin barometer.

Example 5

A mercury barometer has a height reading of 760 mm of mercury. What is the atmospheric pressure in pascals if the density of mercury can be taken as 13 600 kg/m^2 and the acceleration due to gravity 9.81 m/s^2.

Answer

The pressure is given by

$$P = H\varrho g = 0.760 \times 13\ 600 \times 9.81 = 101\ 400 \text{ Pa}$$

Corrections to mercury barometers

The reading given by a mercury barometer needs correcting to allow for a number of effects.

- The brass scale used for the height reading will only be correct at one particular temperature. Readings at all other temperatures will thus be in error. If the scale has been calibrated to be correct at 0°C and the temperature at which the reading is made is θ then

$$H_{\text{true}} = H(1 + \alpha\theta)$$

where H_{true} is the corrected reading of the scale, H is the value read at temperature θ, and α is the linear coefficient of expansion of the metal scale.

- The density of the mercury depends on the temperature. If γ is the coefficient of cubical expansion of the mercury then (see earlier this chapter)

$$\varrho_\theta = \frac{\varrho_0}{1 + \gamma\theta}$$

where ϱ_θ is the density of the mercury at temperature θ and ϱ_0 is the density at 0°C.

- The acceleration due to gravity depends on the geographic latitude (see Table 5.1) and the height above sea level. For each metre above sea level the acceleration due to gravity decreases by 3.1×10^{-6} m/s^2.

- The effect of surface tension on the mercury in the tube is to depress the level (see later this chapter for details).

For comparison purposes it is usual to quote the height of the barometric mercury level in terms of what it would be if the measurement had been made at 0°C. If at temperature θ the true height, i.e. the height allowing for the scale having been calibrated at 0°C, is h_{true}, the mercury density ϱ_θ, and the acceleration due to gravity g then the pressure at that temperature is

$$\text{pressure} = h_{\text{true}}\varrho_\theta g = h(1 + \alpha\theta)\varrho_\theta g$$

where h is the observed height on the scale. The equivalent pressure at 0°C is

$$\text{equivalent pressure} = h_0\varrho_0 g$$

where h_0 is the equivalent height at 0°C and ϱ_0 the density at 0°C. To be equivalent then

$$h_0\varrho_0 g = h(1 + \alpha\theta)\varrho_\theta g$$

$$= \frac{h(1 + \alpha\theta)\varrho_0 g}{1 + \gamma\theta}$$

This can be rearranged as

$$h_0\varrho_0 g = h\varrho_0 g\left[1 - \frac{\{\gamma - \alpha\}\theta}{1 + \gamma\theta}\right]$$

Thus the equivalent pressure is

$$\text{equivalent pressure} = h\varrho_0 g\left[1 - \frac{\{\gamma - \alpha\}\theta}{1 + \gamma\theta}\right]$$

The equivalent reading of the mercury height h_0 is given by

$$h_0 = h\left[1 - \frac{\{\gamma - \alpha\}\theta}{1 + \gamma\theta}\right]$$

It is quite common to quote the atmospheric pressure in terms of the height of the mercury column, e.g. a pressure of 760 mm of mercury. For such readings in one locality to be used for comparison in another the custom is to convert them not only to what they would be at 0°C but also to convert them into the readings that would be given if the acceleration due to gravity was 9.80665 m/s².

$$\text{Corrected reading} = \frac{h_0 g}{9.80665}$$

Example 6

What is the correction that must be applied to a reading of a barometer scale as 765.5 mm at 20°C if the scale is only correct at 0°C? The coefficient of linear expansion of the brass scale is 0.000 0184 /°C.

Answer

Using the expresion given earlier

$$H_{true} = H(1 + \alpha\theta)$$
$$= 765.4(1 + 0.000\ 0184 \times 20) = 765.7 \text{ mm}$$

The correction that has to be applied is + 0.3 mm.

Example 7

If the coefficient of cubical expansion of mercury is 0.000 182 /°C and the coefficient of linear expansion of brass 0.000 018 /°C, what would be the mercury height corrected to 0°C and standard acceleration due to gravity of a measurement made at 24°C of 760.0 mm? Assume that the brass scale was correct at 0°C and that the acceleration due to gravity at the location of the measurement was 9.8121 m/s².

Answer

Using the equation developed earlier

$$h_0 = h\left[\frac{g}{g_s}\right]\left[1 - \frac{\{\gamma - \alpha\}\theta}{1 + \gamma\theta}\right]$$

$$= 0.7600 \times \frac{9.8121}{9.8067}\left[1 - \frac{\{0.000\ 182 - 0.000\ 018\}24}{1 + 1.82 \times 10^{-4} \times 24}\right]$$

$$= 0.7574 \text{ m} = 757.4 \text{ mm}$$

Surface tension effects

Liquids always attempt to reduce their surface area to a minimum. That is why small drops tend to form spheres, a sphere having the smallest surface area for its volume. Large drops can be considered to be spheres which have been squashed out because of the weight of the liquid in the drop. This tendency for the liquid surface to reduce its area to the minimum led to the idea of the surface being rather like an elastic skin which is always trying to contract. The term *surface tension* is due to the idea that this surface 'skin' is in a state of tension, i.e. being held in a stretched state, and it would contract if allowed to do so. Figure 5.12 illustrates a simple

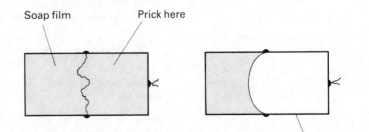

Soap film   Prick here

Fig. 5.12   Soap film in a frame

Wire frame

experiment. A loose thread is tied across a wire frame. The frame is then dipped into soap solution so that a soap film is formed across the entire frame. The thread then lies in the film in a haphazard way. However, if the film is pricked on one side of the thread so that the film breaks, the film on the other side of the thread stretches the thread out into a circular arc.

We can offer an explanation for surface tension in terms of attractive forces between molecules of the liquid. They are always trying to pull the molecules together into the smallest possible shape. A consequence of surface tension is the shape of the meniscus of a liquid in a tube. Figure 5.13 shows the shape of the meniscus at a clean glass surface for water and for mercury. We can explain these shapes in terms of the attractive forces. With water the attractive forces between the glass molecules and the water molecules is, for the water near the glass, larger than that between just water molecules. With mercury the attractive forces between the mercury molecules and the glass molecules is less than that between just mercury molecules.

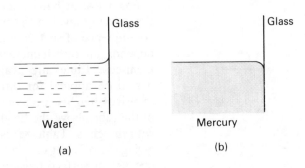

Fig. 5.13 Shape of meniscus, (a) water, (b) mercury

When a tube is dipped into a liquid such as water, the liquid will rise up the tube to an extent which depends on the diameter of the tube, the smaller the diameter the greater the height to which the water rises (Fig. 5.14). If however the tube is dipped into mercury the liquid level inside the tube is depressed, the smaller the diameter the greater the depression. This effect is called *capillarity*. The elevation or depression of a liquid surface for a particular liquid depends on the tube diameter and the shape of the meniscus. Where the shape of the meniscus is that of part of a spherical surface with the same radius as the tube the

$$\text{elevation or depression is proportional to } \frac{1}{\text{tube diameter}}$$

Thus halving the diameter doubles the elevation of depression.

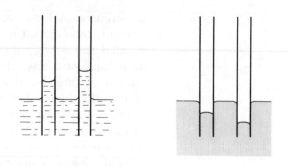

Fig. 5.14 Capillarity, (a) with water, (b) with mercury

(a)　　　　　　(b)

When the meniscus can be considered to be part of a spherical surface which has a radius equal to that of the tube, for clean water in a clean tube the water level rises by about 0.7 mm for a 10 mm diameter tube, about 1.4 mm for a 5 mm diameter tube, about 2.8 mm for a 2.5 mm diameter tube. For clean mercury in a clean glass tube the amount of depression is about 0.4 mm for a 10 mm diameter tube, about 0.8 mm for a 5 mm diameter tube, about 1.6 mm for a 2.5 mm diameter tube.

For a Fortin barometer, i.e. a tube dipping into mercury, the mercury level in the tube is depressed to some extent as a consequence of surface tension. The amount of the depression depends on the diameter of the tube and the shape of the meniscus. Tables are available indicating the correction that has to be applied for different diameter tubes and different mensicus heights, i.e., the height of the mecury in the centre of the tube above the height at which the mercury is in contact with the glass. Table 5.2 is a typical extract from such a table, giving the amount that has to be added to the measured height because of surface tension.

With a U-tube manometer the effect on the liquid levels in the two limbs will cancel out if the two limbs have tubes of the same diameter. If they are of significantly different diameters this will not be the case.

Table 5.2 Surface tension correction

Tube diameter (mm)	Correction to be added to height (mm) for meniscus heights in mm of			
	0.2	0.5	1.0	1.5
5	0.38	0.92	1.62	
10	0.07	0.17	0.32	0.42
15	0.02	0.05	0.09	0.12

Calibration

A simple U-tube manometer needs no calibration since the pressure can be calculated from first principles. Indeed such a manometer can be used as a standard against which other pressure gauges are calibrated. However, the pressure range that can be obtained with a U-tube manometer is fairly low and so other means of calibration have to be used for higher pressures. For this a *dead-weight tester* (Fig. 5.15) is used. The pressure is produced in a fluid by winding in the piston by means of the handle. The pressure so produced is determined by means of adding weights to the platform so that it remains at a constant height. If the total mass of the platform and its weights is M then its weight is Mg. If the cross-sectional area of the platform piston is A then the pressure is Mg/A.

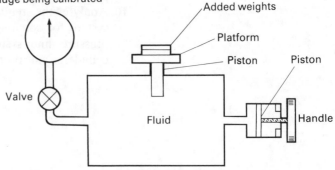

Fig. 5.15 Dead-weight tester

Problems

1 Explain how a U-tube manometer can be used to determine a gas pressure and the types of corrections that will need to be made if an accurate result is required.

2 A U-tube manometer with mercury as the manometer liquid gives a height difference of 120 mm between the mercury level in the two limbs. What is the pressure difference between the gases above the mercury in the two limbs? Take the density of mercury to be 13 600 kg/m³ and the acceleration due to gravity as 9.8 m/s².

3 Explain the principles involved in the measurement of pressure by (a) a diaphragm, (b) a bellows, (c) a Bourdon tube form of instrument and explain how the movement of the sensor can be converted into a display.

4 Explain the principles of operation for mercury column and aneroid barometers.

5 Explain the corrections that may need to be made to the reading obtained from a mercury column barometer for an accurate value of the atmospheric pressure to be obtained.

6 A mercury barometer is found to have a mercury level at 755 mm. What is the atmospheric pressure in pascals. The density of mercury may be taken as 13 600 kg/m³ and the

acceleration due to gravity as 9.81 m/s^2.

7 What is the correction that should be applied to a mercury barometer reading of 750.6 mm at 18°C if the brass scale is only correct at 0°C? The coefficient of linear expansion of brass is 0.000 0184 /°C.

8 A mercury barometer gives a reading of 764.5 mm at a temperature of 21°C. What is (a) the correction that should be applied for the brass scale being accurate only at 0°C, (b) the equivalent pressure at 0°C? The coefficient of linear expansion of brass is 0.000 0184 /°C, the coefficient of cubical expansion of mercury 0.000 182 /°C and its density at 0°C 13 600 kg/m^3. The accleration due to gravity is 9.81 m/s^2.

9 Explain how a dead-weight tester can be used to calibrate a pressure gauge.

10 Analyse the requirements for a pressure measurement system for some specific industrial situation, e.g., the air pressure line system or the pressure of gas stored in cylinders, and propose suitable systems.

6 Measurement of level and density

Level measurement

There are many methods that can be used to determine the level of a liquid. They can be grouped under general headings relating to the principles behind the operation of the sensor. The following are some of the main such groups of methods.

1 *Sight*   These methods involve a direct viewing of the liquid level and include such methods as dipsticks and sight glasses.

2 *Floats*   There are many methods which depend on the use of the movement of a float on the liquid surface or the upthrust force on a float to determine the level. Such methods include floats with cords and pulleys, magnetic float gauges, float-operated potentiometers and torque tubes.

3 *Pressure*   These methods depend on the measurement of a pressure. They may take the form of a measurement of the pressure head due to the height of liquid or the pressure difference between the surface and bottom levels of the liquid, and the bubbler method in which the pressure occurring when bubbles escape from a tube at the bottom of the liquid is measured.

4 *Weight*   The weight of a container plus liquid will depend on the amount of liquid in it and this in turn will depend on the height of liquid. Load cells can be used effectively to determine this weight and so give a measure of the level in the container.

5 *Electrical*   Changes in the level of a liquid can be used to produce changes in the resistance or capacitance of an element standing in the liquid.

6 *Ultrasonics*   The reflection of ultrasonic beams from the surface of a liquid can be used to determine its level.

7 *Radiation*   The blocking off of radiation from a radioactive source can be used to determine the level of a liquid.

Density measurement

Density is defined as the mass per unit volume. Thus if an object has a mass m and volume V it has a density ϱ of

$$\varrho = \frac{m}{V}$$

Many of the methods used for the measurement of level are in fact also sensitive to changes in density. Methods that can be used for density measurement include:

1 *Hydrometers* These are essentially floats, the height of the liquid surface on the float being a measure of the density.
2 *Upthrust measurement* These depend on the measurement of the upthrust acting on an immersed float.
3 *Weight measurement* The weight of a container of liquid depends on the level of the liquid in the container and also its density. For a constant level, weight measurements can give a measure of the density. Load cells can be used for the weight measurement.
4 *Capacitance measurement* The density of the liquid between the plates of a capacitor affects its capacitance.

Sight methods for level measurements

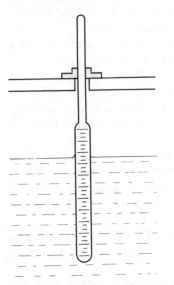

Fig. 6.1 Dipstick

The *dipstick* (Fig. 6.1) is just a metal bar with a scale marked on its side. To determine the level of liquid in a container the dipstick is just held vertically in the liquid. This must always be done from the same fixed position. The stick is then removed and the mark left by the liquid on the stick enables the position of the liquid level to be determined. An obvious example of the use of a dipstick is the determination of the oil level in a car engine. The dipstick is very cheap and fairly accurate, but does not provide a continuous reading.

A *sight glass* (Fig. 6.2) may be just a piece of glass let into the side of the container. The liquid level can then be viewed through the glass. The glass may have a scale marked on it so the level of the liquid can be determined. Such gauges are very widely used in industry. They have the disadvantage of being susceptible to breakage. For this reason they are often used in a branch from the main tank (rather like the tubular level gauge described below), the branch then being capable of being isolated by a valve. They can be used at high pressures and temperatures up to about 500°C.

Another form of sight glass is the *tubular level gauge* (Fig. 6.3). The liquid rises in the tube to the same height as in the container. Because the tube is made of glass or a transparent plastic the level of the liquid can be seen and a reading of the level against a scale obtained. Such gauges are generally limited to pressures not much above atmospheric pressure and temperatures up to about 200°C.

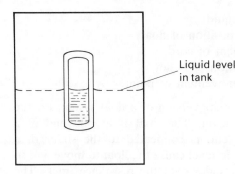

Fig. 6.2 Sight glass

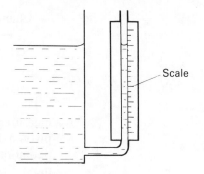

Fig. 6.3 Tubular level gauge

Float methods for level measurement

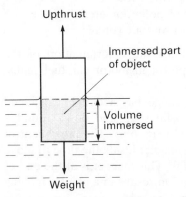

Fig. 6.4 Forces on an object in a fluid

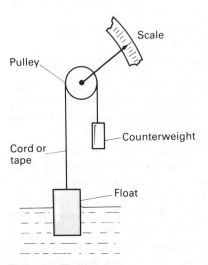

Fig. 6.5 Float gauge

When an object is partially or wholly immersed in a fluid it experiences an upthrust force. The upthrust force is equal to the weight of fluid displaced by the object. This is known as *Archimedes' principle*.

Figure 6.4 shows the forces involved when an object is partially immersed in a liquid. Acting downwards there is the weight of the object. Acting upwards there is the upthrust. The upthrust is determined by the amount of liquid that is displaced by that part of the object below the liquid surface. Thus, for example, if the liquid is water with a density of 1000 kg/m^3 and the volume of the solid below the water surface is 0.0005 m^3 then the mass of water displaced by this volume is 0.0005 × 1000 = 0.5 kg. The weight of this displaced volume is the mass multiplied by the acceleration due to gravity, i.e. 0.5 × 9.8 = 4.9 N. The upthrust is thus 4.9 N.

When an object floats in a liquid there is no net force pushing the object upwards or downwards. Thus the upthrust is then equal to the weight. If the object is then pushed down into the liquid and so displacing more liquid then the upthrust becomes greater than the weight. If the object is partially pulled up out of the water then the part immersed decreases and the upthrust decreases. Then the weight becomes greater than the upthrust.

Figure 6.5 shows one version of a *float gauge* for the measurement of liquid level. The float is attached to one end of a cord or tape which passes over a pulley to a counterweight. When the liquid level rises the float moves upwards and the counterweight pulls the cord taut. In doing this the movement of the cord causes the pulley wheel to rotate. This rotation can be used to move a pointer over a scale. The choice of float material depends on the corrosive nature of the liquid. The most commonly used materials are brass, copper, nickel alloys and stainless steel.

Change in level of liquid
⟶ change in position of float
⟶ movement of cord
⟶ rotation of pulley
⟶ movement of pointer over scale

Figure 3.23 shows another version of a float system for the measurement of liquid level. The float is at one end of a pivoted rod, the other end is connected to the slider of a potentiometer. Changes in level cause the float to move and so move the potentiometer slider over the resistance track. The result is an output voltage which is proportional to the movement of the liquid surface.

Change in level
⟶ change in position of float
⟶ movement of pivoted rod
⟶ movement of potentiometer slider
⟶ change in output voltage

Such a system is widely used for the determination of the level, and hence amount, of petrol in motor car fuel tanks. The method is fairly accurate.

Another version of a float gauge has a doughnut shaped magnetic material float which slides up and down round a sealed tube with the liquid level (Fig. 6.6). The movement of this magnetic material causes a magnet inside the tube to move up and down in sympathy with it. This magnet is at the end of a lever and its up-and-down motion results in a pointer moving across a scale.

Change in level
⟶ change in position of doughnut
⟶ change in position of magnet in tube
⟶ movement of lever
⟶ movement of pointer across scale

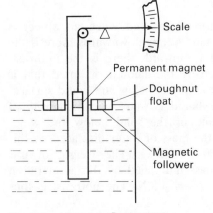

Fig. 6.6 Magnetic float gauge

This arrangement has the advantage that the 'float' arrangement inside the tube does not come into contact with the liquid. There is also no need for leak-proof seals since no seals are required for the float tube.

The above float gauges have all allowed the float to float freely. The *torque tube level gauge* (Fig. 6.7), however, does not do this as the float has to push upwards against a rod. Forces exerted on this rod result in a twisting of a tube. The twist of the tube is monitored by means of strain gauges (see Ch. 2) which thus give a response related to the upthrust. A change in upthrust occurs when there is a change in the weight of liquid displaced by the float. This means when there is a change in the volume of the float immersed, assuming the

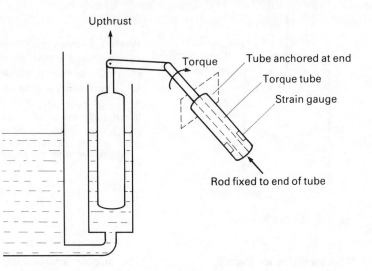

Fig. 6.7 Torque tube level gauge

density of the liquid does not change (see later this chapter). The volume changes when the liquid level changes. Thus changes in liquid level result in changes in the output from the strain gauges.

Change in liquid level
  ⟶ changes in volume of float immersed
    ⟶ changes in weight of liquid displaced
      ⟶ changes in upthrust
        ⟶ changes in twist of tube
          ⟶ changes in resistance of strain gauges

Example 1

A float of volume 0.0002 m³ floats in the surface of water so that half of its volume is below the water line. If the density of water is 1000 kg/m³ what is (a) the upthrust, (b) the weight of the float? Take the acceleration due to gravity to be 9.8 m/s².

Answer

(a) According to Archimedes' principle the upthrust is the weight of fluid displaced. The volume immersed is 0.0002 m³. Therefore the mass of water displaced is 0.0002 × 1000 = 0.2 kg. The weight of water displaced is thus 0.2 × 9.8 = 1.96 N. This is the upthrust.

(b) Because the object is floating the upthrust must be equal to the weight of the float. Hence the weight is 1.96 N.

Example 2

Calculate the change in upthrust acting on a cylindrical float which floats vertically in a liquid of density 1000 kg/m³ and is restrained from moving when the liquid level changes if the level of the liquid rises by 50 mm. The cross-sectional area of the float is 0.020 m². The acceleration due to gravity is 9.8 m/s².

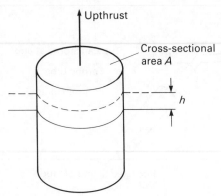

Fig. 6.8 Example 2

Answer

Figure 6.8 shows the situation. The change in upthrust will be equal to the change in the weight of the liquid displaced when the level changes. The change in volume immersed when the level rises by h is hA, where A is the cross-sectional area of the cylinder. The change in weight of liquid displaced is thus $hA\varrho g$, where ϱ is the density of the liquid and g the acceleration due to gravity. The change in upthrust is the change in the weight of fluid displaced. Hence

$$\text{change in upthrust} = hA\varrho g = 0.050 \times 0.02 \times 1000 \times 9.8 = 9.8 \text{ N}$$

Float methods for density measurement

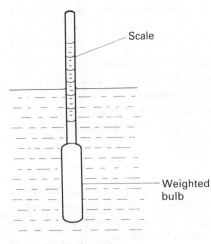

Fig. 6.9 Hydrometer

The simple *hydrometer* (Fig. 6.9) is used to determine the density of liquids. The hydrometer floats in the liquid to a depth which depends on the density of the liquid. A scale on the stem of the instrument enables the density to be read off as the value at the liquid surface.

An object floats when the upthrust acting on it equals its weight. The upthrust, according to Archimedes' principle, is the weight of fluid displaced. This is the volume of the hydrometer below the surface multiplied by the density of the liquid multiplied by the acceleration due to gravity.

$$\text{Weight of hydrometer} = \text{volume immersed} \times \text{density} \times g$$

As the weight of the instrument does not change the volume immersed must change when the density changes. This means that the level of the liquid against the scale on the stem changes. Thus the level becomes a measure of the density of the liquid.

The *torque level tube* described in Fig. 6.7 can be used for the measurement of density. For this purpose the float is completely immersed. Because of this the volume of liquid displaced does not change. The upthrust on the float will thus only change if the density of the liquid changes.

Change in density of liquid
$\longrightarrow$ change in weight of liquid displaced
$\longrightarrow$ change in upthrust
$\longrightarrow$ change in twist of tube
$\longrightarrow$ change in resistance of strain gauges

Example 3

The float of a torque level tube instrument has a volume of 0.005 m^3 and is completely immersed in a liquid. What will be the change in upthrust on the float if the density of the liquid changes from 1000 to 1010 kg/m^3? Take the acceleration due to gravity to be 9.8 m/s^2.

Answer

The upthrust is the weight of liquid displaced by the immersed float. This is the float volume multiplied by the density multiplied by the acceleration due to gravity. Hence initially the upthrust is

initial upthrust $= 0.005 \times 1000 \times 9.8$

and after the density change

new upthrust $= 0.005 \times 1010 \times 9.8$

Thus the change in upthrust is

change in upthrust $= 0.005 \times (1010 - 1000) \times 9.8 = 0.49$ N

Pressure methods for level measurement

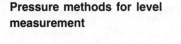

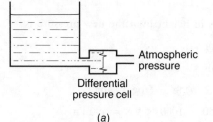

Differential
pressure cell

(a)

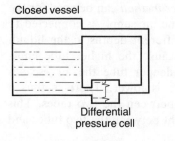

Differential
pressure cell

(b)

Fig. 6.10 Level measurement using a differential pressure cell

The pressure at the base of a tank of liquid due to the liquid is $h\varrho g$, where h is the height of the liquid surface above the base, ϱ the density of the liquid and g the acceleration due to gravity (see Ch. 5). Thus the pressure is proportional to the height h and so is a measure of the level of the liquid in the tank.

Figure 6.10 shows two forms of level measurement based on the measurement of pressure. In Fig. 6.10(a) a diaphragm pressure cell (see Ch. 5) determines the pressure difference between the liquid at the base of the tank and atmospheric pressure when the tank is open to the atmosphere. The pressure difference is $h\varrho g$, where h is the height of the liquid above the base of the tank. With a closed, or open, tank the system illustrated in Fig. 6.10(b) can be used. Here the pressure difference between the liquid at the base of the tank and the air or gas above the surface of the liquid is measured. This equals $h\varrho g$, where h is the height of the liquid surface above the tank base.

Change in level

$\longrightarrow$ change in pressure difference

$\longrightarrow$ change in output from pressure gauge

A different method is the *bubbler method*. This uses a pipe which dips to virtually the bottom of the tank (Fig. 6.11). A constant flow of air, or some other suitable gas, passes into the tube and bubbles out of the bottom. The escaping gas limits the pressure in the equipment which is then proportional to the height of the liquid above the bottom of the tube. The pressure can be measured by means of a diaphragm pressure gauge (see Ch. 5).

Change in level

$\longrightarrow$ change in pressure at which bubbles escape

A major problem with the bubbler method is that gas is introduced into the liquid and this can affect processes involving the liquid. The method does however have the

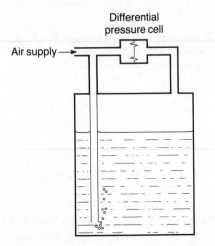

Fig. 6.11 Bubbler method

advantage of only having the tube in the liquid and so if a non-corrosive material is used for the tube the method can be employed with corrosive liquids. It can also be used with slurries.

Example 4

What is the change in pressure produced at the base of tank of water when the level of the water in the tank increases by 50 mm? The density of water is 1000 kg/m^3 and the acceleration due to gravity 9.8 m/s^2.

Answer

The pressure at the base of a tank with water is $H\varrho g$, where H is the height of water, ϱ its density and g the acceleration due to gravity. Thus

initial pressure $= H\varrho g$

and when the water increases in height by h the new pressure is

new pressure $= (H + h)\varrho g$

The change in pressure is thus

change in pressure $= (H + h)\varrho g - H\varrho g = h\varrho g$

$= 0.050 \times 1000 \times 9.8 = 490\,\text{Pa}$

Pressure method for density measurement

Figure 6.12 shows how the *bubbler method* can be used for the measurement of density. The same air supply is connected to two tubes. These tubes are at different depths in the liquid. Bubbles emerge from them. Because the higher tube emits bubbles at less depth than the deeper tube the air escapes more easily and so the pressure is lower. There is thus a difference in pressure produced between the two tubes. This depends on the difference in height between the two tubes and

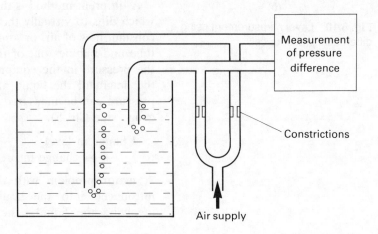

Fig. 6.12 Bubbler method for density measurement

also the density of the liquid ($P = h\varrho g$). Thus for a constant height difference between the tubes the pressure difference is related to the liquid density. Hence a measurement of the pressure difference is a measure of the density.

Weight methods for level measurement

Figure 6.13 illustrates the principle involved in the use of load cells (see Ch. 2) to determine liquid level in a container. Essentially the load cells, which are included in the supports for the container, determine the container weight. Since the weight depends on the level of liquid in the container then the load cells give responses related to liquid level. The load cells deform under the action of the liquid weight and the deformation can be monitored by strain gauges attached to the walls of the load cell. The gauges can be incorporated in a Wheatstone bridge (see Ch. 3) and the out-of-balance voltage then becomes a measure of the liquid level.

Change in liquid level
$\longrightarrow$ change in weight of container plus liquid
$\longrightarrow$ change in forces acting on container supports
$\longrightarrow$ change in forces on load cells
$\longrightarrow$ deformation of load cells
$\longrightarrow$ change in strain for strain gauges
$\longrightarrow$ change in resistance of strain gauges
$\longrightarrow$ out-of-balance voltage in bridge

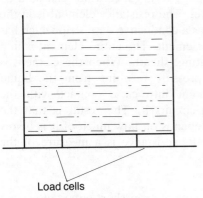

Load cells

Fig. 6.13 Load cell system

Since the load cell is not in the liquid the method can be used for corrosive liquids with no problems. Problems can occur if the tanks containing the liquid are subject to sideways forces, perhaps from the wind. This can give rise to errors in the readings obtained. The method can be used with liquids, slurries and solids.

Weight methods for density measurement

If the level of a liquid in a container is kept constant then any change in its weight will be due to a change in density.

Weight $= mg = V\varrho g$

where m is the mass of the liquid, ϱ its density and V its volume with g the acceleration due to gravity. The weight of a container can be monitored by means of a load cell, as described in Fig. 6.13. Hence changes in density will result in weight changes which in turn increase the load on a load cell. The resulting deformation of the load cell can be monitored by means of strain gauges. Hence the resulting change in resistance of the strain gauges becomes a measure of density changes in the liquid.

Electrical methods for level measurement

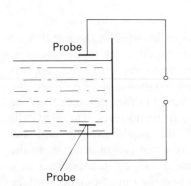

Probe

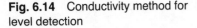

Probe

Fig. 6.14 Conductivity method for level detection

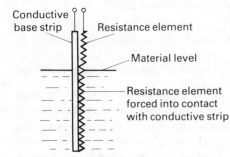

Conductive base strip

Resistance element

Material level

Resistance element forced into contact with conductive strip

Fig. 6.15 Resistance level gauge

Electrical methods for density measurement

Conductivity methods can be used to indicate when the level of a liquid in a container reaches a critical level. Figure 6.14 shows the basic principle. One probe is mounted in the liquid and the other at the required level. When the liquid is short of the required level the resistance between the two probes is high since part of the path between the probes is air. If the liquids are reasonable conductors of electricity then when the liquid level reaches the level of the air probe the resistance between the probes drops by quite a large amount. This drop in resistance can be detected. The arrangement is thus a detector for when the liquid reaches a particular level. Such a method is used with high-conductivity water-based materials, e.g. milk, beer, wines, soups, etc.

Figure 6.15 shows a *resistance level gauge* which can be used for the measurement of level. The resistance element is in the form of a strip. The strip has a conductive base strip which has close to it a flat resistance element. The entire element is in a protective, electrical insulating, sheath. When the element is vertically in a liquid the pressure resulting from the liquid forces the resistance element into close contact with the conductive strip and short-circuits it. The result is that the resistance of the element depends on how much of it is below the liquid surface. Hence the resistance is a measure of the level.

Change in level
$\longrightarrow$ change in amount of strip short-circuited
$\longrightarrow$ change in resistance of strip

Figure 2.9 shows one form of a *capacitive method* for the determination of liquid level. The capacitor consists of two concentric cylinders with the liquid between them. When the liquid level changes the capacitance changes (see Ch. 2). The probe must be insulated. Errors will arise as a result of temperature changes since a change will produce a change in capacitance without any change in level occurring. Errors might also be produced if the electrodes become coated with materials from the liquid.

The concentric cylinder capacitor, described above and in Fig. 2.9, can be used for the determination of density. If the capacitor is completely immersed so that there is no change in the level of the liquid between the cylinders then a change in the density of the liquid produces a change in capacitance. Hence the measurement of the capacitance gives a measure of the density.

Ultrasonic methods for level measurement

Ultrasonic transmitter and receiver

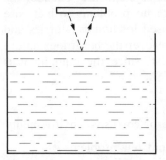

Fig. 6.16 Ultrasonic level gauge

Radiation methods for level measurement

In one version of an ultrasonic level gauge an ultrasonic transmitter is placed above the surface of the liquid and emits pulses of ultrasonics (Fig. 6.16). The pulses are reflected from the surface to a receiver. The time taken from emission to reception of the reflected pulse can be measured. Since the time taken depends on the distance of the liquid surface from the transmitter/receiver the level of the liquid can be determined.

Changes in level
⟶ changes in the time taken for a pulse to travel to the liquid surface and back to the receiver

Figure 6.17 shows two forms of *radiation level measurement systems*. Radiations are emitted from a radioactive source and pass through the container walls and any intervening liquid before reaching a nuclear radiation detector. Radiation is absorbed by the liquid and so the presence of liquid between source and detector reduces the radiation detected. Figure 6.17(*a*) shows the arrangement with a compact source and an extended detector. This arrangement can be used for the measurement of level over the length of the detector. Figure 6.17(*b*) shows the situation when there is a compact source

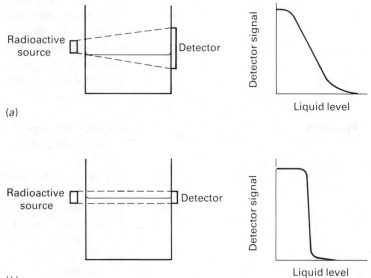

Fig. 6.17 Radiation level measurement systems

and a compact detector. With such an arrangement the output from the detector is quite sensitive to changes in level over a small range.

Since all the elements of the system are outside the container and its liquid, there are no problems with its use with corrosive liquids. It can be used for liquids, slurries and solids. The major problems are however the hazardous nature of the radioactive source and cost, since the equipment is relatively expensive.

Level measurements in corrosive environments

For level measurements in hostile or corrosive environments:

- sight glasses are likely to present few problems
- floats may present problems if the liquids can coat the floats and so change the buoyancy, also many of the systems employ seals and these may require frequent maintenance in such environments
- diaphragm pressure cells can use corrosion-resistant materials for casings and diaphragms
- with bubblers only the tube has to be of corrosion-resistant materials
- weight measurement by load cells presents no problems since the load cell is not in contact with the liquid
- with capacitance measurement systems the capacitor plates may have to be coated with an insulating, inert, plastic both to avoid corrosion and also limit any electric currents that may occur between the plates
- ultrasonic measurement systems in which there is no physical contact with the liquid present no direct problems of corrosion though the accuracy may be affected by vapours in the intervening space between the transmitter/receiver and the liquid surface
- radiation methods can involve no physical contacts between the equipment and the liquid and so the nature of the fluid will present no problems

Problems

1 Explain the operating principles of level measurement systems based on:
   (a) floats,
   (b) pressure measurement between the top and bottom of a container of liquid,
   (c) radiation from a radioactive source.
2 Explain the operating principles of the following density measurement systems:
   (a) the hydrometer,
   (b) bubblers,

(c) load cells.

3 What are the problems associated with the use of level measurement systems based on floats for corrosive liquids?

4 A float is in the form of a cylinder with its axis vertical in a liquid. The cylinder has a cross-sectional area of 0.010 m^2 and a length of 0.40 m. The acceleration due to gravity is 9.8 m/s^2.
 (a) If the cylinder floats in water, density 1000 kg/m^3, with half its length immersed, what is the weight of the cylinder?
 (b) What is the upthrust force acting on the cylinder when it is completely immersed?

5 A pressure measurement system is to be used for the measurement of the level of a liquid in a tank. What will be the sensitivity required of the pressure gauge if the smallest change in level that has to be detected is 100 mm. The liquid has a density of 1200 kg/m^3 and the acceleration due to gravity is 9.8 m/s^2.

6 Suggest, giving reasons for your choice, level measurement systems that could be used for the following situations:
 (a) to indicate when milk in a container has reached a particular level,
 (b) for use with a highly corrosive liquid,
 (c) to operate a valve closing off water entering a tank when the required level is reached without any need for electrical power supply or pressurised air.

7 What factor determines the choice of length of the detector in a radiation level measurement system which has a compact radioactive source?

8 In the brewing industry a liquid, known as wort, is boiled with hops. A level indicator is required for the mixture. Propose one, bearing in mind that there can be froth and the liquid is hot.

9 (a) Design a float system that could be used to give a display indicating the water level in a container. The display needs to indicate the water level to an accuracy of about 1 cm and a range of 30 cm.
 (b) Set up your system. Test and calibrate it.

7 Measurement of flow

Fluid flow

The term *fluid* is used to describe substances that flow. It thus refers to both liquids and gases. Figure 7.1(*a*) shows the flow of an ideal fluid through a pipe. Every particle of the fluid moves in straight lines parallel to the tube walls with the same velocity. Such behaviour is termed ideal because real fluids do not behave like that. This is because of what is termed *viscosity*.

When a real fluid flows past a solid boundary the fluid immediately adjacent to the solid is slowed down. We can think of there being 'frictional' forces between the solid and the liquid which slow the fluid down. Instead of the term frictional, however, we use the term viscous forces. Because viscous forces slow down fluid motion the term *viscous drag* is frequently used. A consequence of these forces is that with slow fluid flow through a tube the picture is like that shown in Fig. 7.1(*b*). Only in the centre of the tube does the flow reach full velocity. At the tube walls the velocity of the fluid is right down to zero. The term *laminar flow* or *streamline flow* is used for the type of flow shown in Fig. 7.1(*b*). This is because particles of the fluid move along in lines (layers) parallel to the tube walls. They all move in straight lines parallel to the tube walls with velocities which depend on their distance from the tube walls.

At higher rates of flow the motion becomes chaotic. Particles of fluid are found to move in all sorts of directions. If we could follow one particle it would be found to follow a very irregular path. Such motion is said to be *turbulent flow*. Despite the chaotic nature of the flow the average result for flow along a tube is like that shown in Fig. 7.1(*c*). There is after all still a flow of fluid through the tube.

A consequence of viscosity is that when a fluid flows through a pipe there is a drop in pressure along the length of the pipe (Fig. 7.2). This represents the energy used to drive the fluid

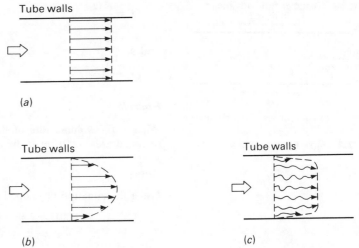

Fig. 7.1 The velocities in the flowing fluid when it is (*a*) an ideal fluid, (*b*) a real fluid with laminar flow, (*c*) a real fluid with turbulent flow

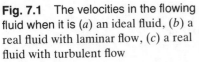

Fig. 7.2 Pressure drop along a pipe

through the pipe against the viscous drag forces.

It is much easier to pour water through a pipe than a thick oil. We can explain this by water having a higher viscosity than the oil (Fig. 7.3). A higher viscosity means that there are much greater viscous forces slowing down the fluid motion through the pipe. However we can get the oil to flow more easily through the pipe if we raise its temperature. Viscosity generally decreases when the temperature is increased.

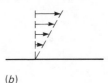

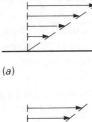

Fig. 7.3 Comparative laminar flow with fluids having (*a*) low viscosity, (*b*) high viscosity, (*c*) high viscosity but now with higher temperature

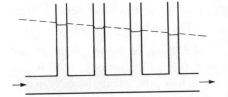

Flow through a tube

Despite velocity through a tube being different at different distances from the tube walls we can consider the flow through the tube in terms of an average velocity. Thus if the average velocity is v then in a time t the flow will have advanced a distance vt (Fig. 7.4). If the cross-sectional area of the tube is A then the volume of fluid that has moved through this distance in time t is Avt. The volume rate of flow Q is the rate at which a volume of fluid flows through the pipe. Hence

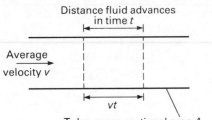

Fig. 7.4 Flow through a pipe

$$Q = \frac{Avt}{t}$$

and so

$$Q = Av$$

Example 1

What is the volume rate of flow of water through a pipe of cross-sectional area 0.001 m^3 if the average velocity of flow is 2 m/s?

Answer

Using the equation developed above

$$Q = Av = 0.001 \times 2 = 0.002 \text{ m}^3/\text{s}$$

Types of measurements

The term *flow measurement* can be considered to include measurements of three different quantities, and hence different measurement techniques. The quantities measured are:

1 *The actual velocity of the fluid at some point in the fluid* The velocity through a tube, for example, is different in the central part of the tube from that near the tube walls. The unit is m/s. The Pitot static tube is used for such a measurement.

2 *The volume rate of flow* The rate at which a volume of a fluid passes through a pipe; the unit is m^3/s. Such measurements are made using tubes, like the Venturi tube, which produce a pressure difference as a result of the flow being made to flow through a constriction, turbine flowmeters and positive displacement meters.

3 *The mass rate of flow* The rate at which a mass of fluid passes through a pipe, the unit is kg/s. The mass flow rate is usually deduced from the volume flow rate, being the volume flow rate multiplied by the fluid density. Some specialised techniques are, however, available for its direct measurement.

Pitot static tube

The *Pitot static tube* (Fig. 7.5) is used for velocity measurements, the pressure difference being measured between a point in the fluid where the fluid is in full flow and a point at rest in the fluid. The difference in pressure is due to the kinetic energy of the fluid and so proportional to the square of the velocity. This is the relationship for a liquid. For a gas the relationship generally needs modification. The pressure difference is often measured with a diaphragm form of pressure gauge. Pitot static tubes can be used for measurements of fluid velocities as low as 1 m/s and as high as 60 m/s and for both liquid and gas flow. There is always the problem with the Pitot static tube that the holes will become blocked. For this reason it is mainly used with gases rather than liquids.

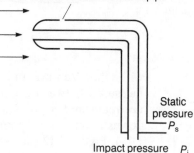

Holes around circumference of pipe

Static pressure P_s

Impact pressure P_i

Fig. 7.5   Pitot static tube

Constriction flowmeter

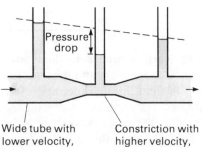

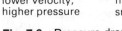Pressure drop

Wide tube with lower velocity, higher pressure

Constriction with higher velocity, smaller pressure

Fig. 7.6   Pressure drop at a constriction

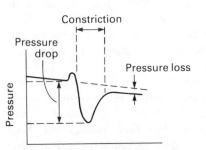

Constriction

Pressure drop

Pressure loss

Pressure

Distance along tube

Fig. 7.7   Pressure changes with a constriction flowmeter

This form of flowmeter is used for the measurement of the volume rate of flow. It is based on the principle that a pressure drop occurs at a constriction in a pipe, as in Fig. 7.6 (see Ch. 2). Figure 7.7 shows how the pressure might vary along a tube in which there is such a flowmeter. The amount by which the pressure drops depends on the volume rate of flow Q.

Q is proportional to $\sqrt{}$(pressure difference)

After the constriction the pressure does not resume completely the pressure that occurred before the constriction. There is some pressure loss. The amount of loss depends on the form of the constriction.

A principle of fluid flow can be stated as: when a fluid is caused to accelerate, i.e. increase its velocity, the fluid pressure drops. The constriction form of flowmeter is an example of this. Another example is the lift force experienced by an aerofoil (Fig. 7.8). When the fluid flows over the surfaces of the aerofoil its shape is such that the flow over the upper surface is at a higher velocity than that over its lower surface. The consequence of this is that the pressure above the aerofoil is less than that below it. Hence there is a net upward force.

There are a number of forms of flowmeter based on this measurement of the pressure difference between the flow in the full cross-section tube and the constriction. They differ in the way they produce the constriction. With the *Venturi tube* (Fig. 7.9) there is a gradual tapering of the pipe from the full diameter to the constricted diameter. The pressure difference between the flow prior to the constriction and at the constriction can be measured with a simple U-tube manometer or a diaphragm pressure cell. The *orifice flowmeter* (Fig. 7.10) is simply a disc, with a central hole, which is placed in the tube. This results in a similar flow pattern to that occurring

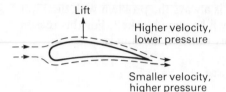

Fig. 7.8 Lift with an aerofoil

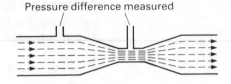

Fig. 7.9 Venturi tube

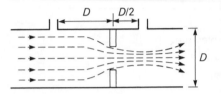

Fig. 7.10 Orifice plate

Fig. 7.11 Orifice plates

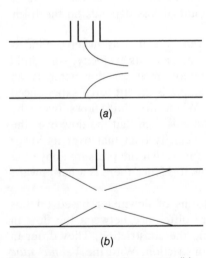

Fig. 7.12 (a) Nozzle flowmeter, (b) Dall flowmeter

with the Venturi tube. The pressure difference can be measured between a point equal to the diameter of the tube upstream and a point a distance equal to half the diameter downstream, or at points on either side of the plate (Fig. 7.11). Figure 7.12 shows two further versions of flowmeter, the *nozzle flowmeter* and the *Dall flowmeter*.

The Venturi tube offers the least resistance to fluid flow, giving the smallest pressure loss (see Fig. 7.7) and thus has the least effect on the rate of flow of fluid through the pipe while the orifice plate offers the greatest resistance. The greater the pressure loss due to the flowmeter the more power has to be expended in pumping the fluid through the pipe. All the flowmeters have long life without maintenance or recalibration being required and an accuracy of about ± 0.5% for all but the orifice plate which is likely to have an accuracy of ± 1.5%. The orifice plate gives the greatest pressure loss and the worst accuracy but is the cheapest. The Venturi tube can cope with dilute slurries but the others tend to require fluids which contain no particles. This is because deposits of particles can occur in regions behind plates, i.e. silting, and affect the operation of the flowmeter.

Another form of constriction flowmeter is the *variable area flowmeter*. A common form of variable area flowmeter is the *rotameter* (Fig. 7.13). This employs a float in a tapered vertical tube. The fluid in flowing up the tube pushes the float upwards. But the fluid has to flow through the constriction which is the gap between the float and the walls of the tube. The result is a pressure drop. The tube is tapered and the gap between the float and the tube walls increases as the float moves up the tube. This means that the pressure drop becomes smaller the further the float moves up the tube. The float moves up the tube until the fluid pressure is just sufficient

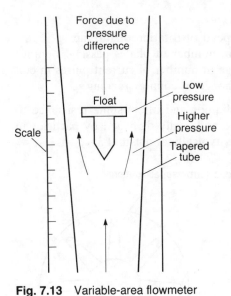

Fig. 7.13 Variable-area flowmeter

to balance the weight of the float. The greater the flow rate the greater the pressure difference for a particular gap. The float thus moves up the tube to a height which depends on the rate of flow. A scale alongside the tube can thus be calibrated to read directly the flow rate corresponding to a particular height of float. The rotameter can be used to measure flow rates from about 30 ml/s (30×10^{-6} m^3/s) to 120 l/s (120×10^{-3} m^3/s). It is a relatively cheap instrument, capable of a long life with little maintenance or recalibration being required. It is not highly accurate, about $\pm$ 1%, and has a significant effect on the flow.

Example 2

How will the pressure difference measured with a Venturi tube change if the volume rate of flow is doubled?

Answer

Since the volume rate of flow is proportional to the square root of the pressure drop then

(volume rate of flow)^2 is proportional to pressure drop

Hence doubling the rate of flow increases the pressure drop by a factor of 2^2, i.e. four.

Example 3

What are the advantages and disadvantages of the orifice plate form of constriction flowmeter in comparison with other forms of constriction flowmeter?

Answer

The main advantage is that it is cheap. The disadvantages are that it produces a large pressure drop, it is more inaccurate, and can only be used with fluids containing no particles.

Turbine flowmeter

The *turbine flowmeter* operates on the same principle as a child's toy windmill on the end of a stick. The faster the flow of air through the windmill the faster it rotates. The turbine flowmeter (Fig. 7.14) consists of a multi-bladed rotor that is supported centrally in the pipe along which the flow occurs and which rotates as a result of the fluid flow. The angular rate at which the blades rotate is proportional to the flow rate. The rate of revolution of the rotor can be determined using a magnetic pick-up. The blades are made of a magnetic material and every time they pass a coil a current pulse is produced in the coil as a result of electromagnetic induction. The pulses are counted and so the number of revolutions of the rotor determined.

Change in rate of flow
⟶ change in speed of rotation of turbine
⟶ change in number of blades passing a point
⟶ change in number of current pulses in coil
⟶ change in counter reading

The meter is used with liquids and offers some resistance to fluid flow. It is expensive and easily damaged by particles in the fluid. The accuracy is typically about ± 0.3%.

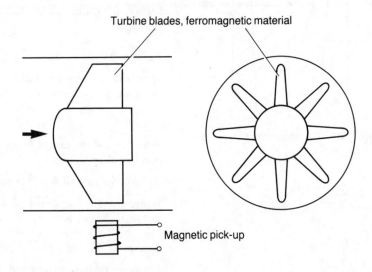

Fig. 7.14 Turbine meter

Positive displacement meters

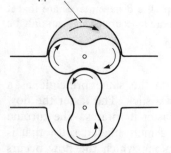

Fig. 7.15 Rotating lobe meter

This form of flowmeter works on the principle of dividing up the flowing fluid into known volume packets and then counting them to give the total volume passing through the meter. They are widely used for water meters, gas meters and petrol-pump meters to determine the volume delivered. If the volume delivered over a particular time is monitored then the volume rate of flow can be established. There are many forms of positive displacement meter, e.g. rotating lobe meter, rotating vane meter, nutating disc meter, reciprocating piston meter. Some are used with liquids, others with liquids. Figure 7.15 shows the *rotating lobe meter*, this being used with gases. Each time the lobes rotate they trap a volume of the fluid and move from the pipe on the left to the one on the right. Thus the amount of fluid which passes through the meter is determined by the rate at which the lobes rotate. Accuracies are generally of the order of ± 0.2%.

Calibration

Calibration of flowmeters for use with liquids can be carried out by measuring the volume, or mass, of liquid passed through a pipe in a measured time (Fig. 7.16). An alternative,

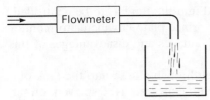

Fig. 7.16 Calibration of a flowmeter

with gases or liquids, is to calibrate a flowmeter against some form of standardised flowmeter.

Example 4

The mass of water passing through a flowmeter in 400 s was measured as being 36.08 kg. If the density of water is 1000 kg/m³ at the temperature at which the measurement was made, what is the average rate of flow?

Answer

The volume of water passing through the flowmeter in 400 s is given by using the equation

$$\text{density} = \frac{\text{mass}}{\text{volume}}$$

$$\text{Volume} = \frac{\text{mass}}{\text{density}} = \frac{36.08}{1000} = 0.03608 \text{ m}^3$$

Hence the volume rate of flow is

$$\text{volume rate of flow} = \frac{0.03608}{400} = 0.000\ 0902 \text{ m}^3/\text{s}$$

Problems

1 Explain the the following terms:
   (a) fluid,
   (b) laminar flow,
   (c) viscous drag.
2 Explain the principle of operation of constriction flow-meters.
3 Explain the basic principles involved in the operation of the following flowmeters:
   (a) Pitot static tube,
   (b) Orifice plate,
   (c) Rotameter,
   (d) Turbine.
4 Suggest a flowmeter that could be used for the measurement of the volume flowrate of a liquid in which there may be some fine particles.
5 A fluid of density 1200 kg/m³ moves through a pipe of cross-sectional area 0.010 m² with an average velocity of 2.0 m/s. What is (a) the volume rate of flow, (b) the mass rate of flow?
6 In a calibration test of a flowmeter 5.0 kg of a liquid of density 1100 kg/m³ is determined as flowing through the meter in 500 s. What is the average volume rate of flow?
7 In the brewing industry ground malt is mixed with hot water to form a mash. This then flows into a mash tun. The

rate of flow of this liquid into the tun has to be controlled. It is proposed to use an orifice plate flowmeter to monitor the flow. Discuss the advantages and disadvantages of this proposal.

8 Petrol pumps deliver fuel through a hose into the tank of a car. The amount of fuel delivered is displayed on an indicator. Suggest a flowmeter that could be used for this purpose.

8 Measurement of temperature

Temperature

Temperature can be considered to be a measure of the degree of hotness of an object. When two objects at the same temperature are put into contact with each other, their temperatures do not change.

Object at temperature θ_1	no net flow of energy between them	Object at temperature θ_1

This is true whatever the size or form of the two objects. When two objects at different temperatures are put into contact with each other then energy flows between them until the two come to the same temperature. Thus if temperature θ_1 is greater than θ_2,

Object at temperature θ_1	net flow of energy $\longrightarrow$	Object at lower temperature θ_2

This is the basis of the measurement of temperatures, i.e. the basis of thermometers. When a thermometer is put into contact with an object then energy flows between the two until the thermometer is at the same temperature as the object.

Temperature scales

A scale, for any quantity, can be established by defining two points and then deciding how the interval between the points is to be divided up. Thus we could specify a scale for length by taking a piece of wood and stating that one end of it is to be called length 0 and the other end length 100. The interval between we could divide into 100 equal parts. We could thus mark the piece of wood and by comparison with it determine the lengths of other objects. We can similarly specify a scale for temperature by taking a thermometer and stating that its reading when at the temperature of melting ice is to be given the value 0 and at the temperature of boiling water 100. The interval between could be divided into 100 equal parts. We can then determine the temperatures of other objects by comparison with our standard thermometer. This is a scale that is used, being essentially what we call the *Celsius scale*. Each of the 100 equal parts is called a Celsius degree (°C).

In defining a temperature scale we need to specify two fixed points, how the interval between them is to be subdivided, and what thermometer we will use as a standard for temperatures between those points. The fixed points need to be ones which we are certain can be reproduced time and time again and in different places and still be the same. For this reason freezing points and boiling points of pure substances are chosen. The thermometer to be specified as standard must be one which will accurately give the same answers whenever and wherever it is used.

The temperature scale used world-wide is the *International Practical Temperature Scale*. This uses a number of fixed points and specifies what thermometers are to be used to determine the temperatures in the intervals between the fixed points. Temperatures on this scale can be expressed in two different way, in Celsius degrees (°C) or kelvins (K). The relationship between these temperatures is

Temperature in K = temperature in °C + 273.15

Table 8.1 shows the fixed points and Table 8.2 the methods to be used to determine temperatures between them. These well-defined temperatures and methods of obtaining intermediate temperatures (the term used is *interpolation*) enable laboratories readily to calibrate thermometers.

Example 1

What is the temperature on the kelvin scale of (a) 20°C, (b) 500°C, (c) −30°C?

Answer

The relationship between temperatures on the kelvin scale and those on the Celsius scale is

Temperature in K = temperature in °C + 273.15

Because no decimal figures are quoted for the Celsius temperatures it is adequate to take the conversion figure as 273. Hence:

(a) Temperature in K = 20 + 273 = 293 K

(b) Temperature in K = 500 + 273 = 773 K

(c) Temperature in K = − 30 + 273 = 243 K

Table 8.1 The International Practical Temperature Scale fixed points

Fixed point	Temperature	
	°C	K
Triple point of hydrogen	−259.34	13.81
Boiling point of hydrogen at 33 330.6 Pa pressure	−255.478	17.042
Boiling point of hydrogen	−252.24	20.28
Boiling point of neon	−246.048	27.102
Triple point of oxygen	−218.789	54.361
Triple point of argon	−193.352	83.798
Boiling point of oxygen	−186.962	90.188
Triple point of water	0.01	273.16
Boiling point of water	100	373.15
Freezing point of tin	231.9681	505.1181
Freezing point of zinc	419.58	692.73
Freezing point of gold	1064.43	1337.58

Note: The above are termed the *primary fixed points*. There are also secondary fixed points to aid in specifying temperatures within the intervals between the primary fixed points. Unless otherwise stated the boiling points are at normal atmospheric pressure, i.e. 101 325 Pa. The triple point is the temperature at which the solid, liquid and gas coexist.

Table 8.2 The International Practical Temperature Scale thermometers

Temperature range*		Standard interpolation thermometer
°C	K	
−259.34 to 0	13.81 to 273.15	Platinum resistance
0 to 630.74	273.15 to 903.89	Platinum resistance
630.74 to 1064.43	903.89 to 1337.58	Platinum–10% rhodium/ platinum thermocouple
Above 1064.43	Above 1337.58	Radiation pyrometer

Note: With the platinum resistance thermometer different equations are used to specify temperatures between the various sets of fixed points within the specified ranges.

Methods of measurement

Changes in temperatures lead to changes in some of the properties of materials. These changes can then be used as a measure of temperature. Changes that are commonly used are:

1 *Expansion* An increase in temperature results in materials expanding. The expansion of solids is used in bimetallic-strip thermometers, the expansion of liquids in the mercury-in-glass thermometer, the expansion of gases in the gas thermometer.

2 *Change of state* When the temperature increases more of a liquid will evaporate. This is the basis of the vapour pressure thermometer.

3 *Change in resistance* When the temperature increases the resistance of a metal wire or strip of semiconductor changes. This change in resistance is the basis of the resistance thermometer.

4 *Production of an e.m.f.* When the temperature of a junction between two dissimilar metals is increased then the potential difference across the junction increases. This is the basis of the thermocouple.

5 *Change in radiation* The higher the temperature of an object the more radiation it emits. Measurement of the amount of radiation is the basis of the radiation pyrometer. Not only does the amount of radiation increase when the temperature increases but so also does its colour. Thus, for example, an increase in temperature may mean an object turning from red hot to white hot. A measure of this colour is the basis of the disappearing filament pyrometer.

Bimetallic strips

An increase in temperature of a strip of metal results in the strip increasing in length. The amount by which the strip expands depends on:

- the amount by which the temperature changes; the greater the change in temperature the greater the expansion;
- the initial length of the strip; the longer the initial length of the strip the greater the expansion;
- the metal concerned.

The above factors can be combined in an equation

$$\text{change in length} = L_0 \theta \alpha$$

where L_0 is the initial length, θ the change in temperature and α a constant for a particular metal, being called the *coefficient of linear expansion*.

The *bimetallic strip* consists of two different metal strips of the same length bonded together (Fig. 8.1). The metals have different coefficients of expansion. When the temperature increases both strips need to expand. But one strip needs to expand more than the other. Because the two strips are bonded together, the only way this can happen is if the strip curves. The strip on the outside of the curve will then have a greater length than the one on the inside of the curve. The amount by which the strip curves depends on:

1 *The two metals used* A high-coefficient-of-expansion metal is combined with a low-coefficient-of-expansion metal

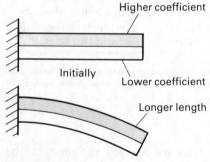

Fig. 8.1 Bimetallic strip

since the greater the difference between their coefficients the greater the amount by which the strip curves.

2   *The length of the composite strip*   The longer the length of the strips in the composite the more each needs to expand and so the greater the amount by which the composite strip curves.

3   *The change in temperature*   The greater the increase in temperature the more each of the strips needs to expand and so the greater the amount by which the composite strip curves.

If one end of a bimetallic strip is fixed the amount by which the other end moves is a measure of the temperature of the strip. This movement may be used as a temperature-controlled switch, as in the simple thermostat used with many domestic central-heating systems (Fig. 8.2). The gap between the free end of the strip and an electrical contact is adjusted so that at the required temperature the strip curves just enough to make contact with the electrical contact. The result is to allow a current to flow in an electrical circuit.

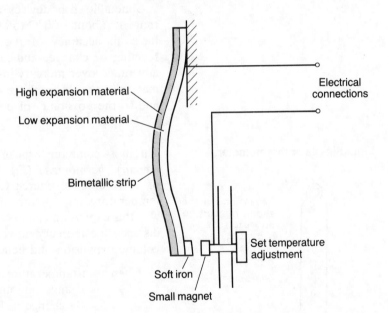

High expansion material

Low expansion material

Bimetallic strip

Electrical connections

Set temperature adjustment

Soft iron

Small magnet

Fig. 8.2   Bimetallic thermostat

The amount by which a bimetallic strip curves depends on its length, the longer its length the greater the amount by which it curves. One way of having a long length of bimetallic strip in a compact form is to wind the strip into a spiral or a helix. If we fix one end of such a strip then when the temperature changes the movement of the free end of the strip, can be used directly to move a pointer across a scale (Fig. 8.3).

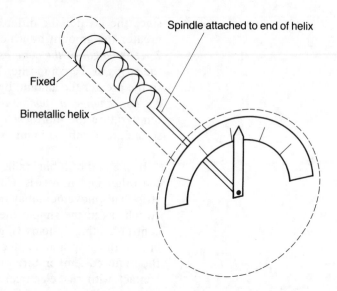

Fig. 8.3 Bimetallic thermometer

Bimetallic-strip devices are robust, can be used within a range of about −30°C to 600°C, can be used for thermostats, have an accuracy of the order of ± 1%, are fairly slow reacting to change, and are relatively cheap. They have an advantage over mercury-in-glass or mercury-in-metal thermometers in that a breakage does not result in mercury leaking and so the possibility of poisonous fumes.

Liquid-in-glass thermometers

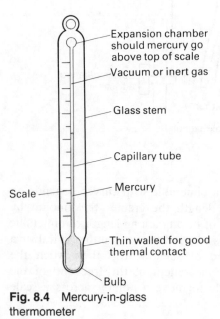

Fig. 8.4 Mercury-in-glass thermometer

The most common form of liquid-in-glass thermometer is the *mercury thermometer* (Fig. 8.4). This relies on the expansion of a volume of mercury. When there is an increase in temperature the volume of the mercury increases. The only way this expansion can take place is up the capillary tube. The distance the mercury moves up the tube is a measure of the volume expansion and hence a measure of the temperature.

Change in temperature
  ⟶ change in volume of the mercury
    ⟶ change in level of the mercury in the tube

The liquid-in-glass thermometer is direct reading, fragile, capable of reasonable accuracy under standardised conditions, fairly slow reacting to change and cheap. With mercury as the liquid the range is −35°C to 600°C, with alcohol −80°C to 70°C, with pentane −200°C to 30°C. Thermometers are calibrated for use in three different ways:

1 *Partially immersed* The thermometer is intended to be used with a hot liquid so that the level of the liquid comes up to a particular mark on the thermometer stem.

2 *Totally immersed* The thermometer is to be used immersed to the level of the liquid in the thermometer stem.

3 *Completely immersed* The thermometer is to be completely immersed.

If a thermometer is not immersed to the amount for which it was calibrated then errors occur.

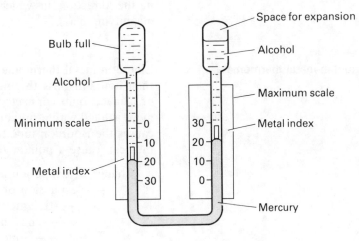

Fig. 8.5 Maximum and minimum thermometer

There are a number of special forms of liquid-in-glass thermometers. Figure 8.5 shows the *maximum and minimum thermometer*. This is used to indicate the maximum and minimum temperatures that occur during the time elapsed since the instrument was set. The thermometric liquid used is alcohol. When the alcohol in the left-hand limb expands it pushes the mercury thread. Resting on the mercury surface in the left hand limb is a small metal index. When the mercury level is pushed by the expanding alcohol the index is left behind. It indicates the highest position of the mercury in the left-hand limb. This is the smallest volume of the alcohol and so is the minimum temperature. The movement of the mercury in the right-hand limb pushes an index ahead of it. The greater the amount of expansion of the alcohol in the left-hand limb the further up the right-hand limb the index is pushed. Its position thus indicates the maximum temperature reached.

Another special form of mercury-in-glass thermometer is the *clinical thermometer* (Fig. 8.6). This has a narrow constriction in the capillary tube just above the bulb. When the thermometer is used to measure the temperature of a patient and the temperature increases, the mercury forces its way through the constriction. However, when the thermometer is taken from the patient and the temperature drops, the mercury above the constriction cannot force its way back

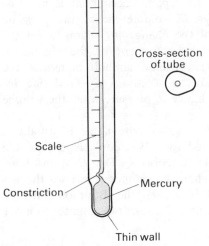

Fig. 8.6 Clinical thermometer

through the constriction and thus the temperature of the patient can be read. The mercury can be returned to the bulb by shaking the thermometer. The thermometer has a restricted temperature scale (35–42°C, with 37°C being the healthy-patient temperature) just covering the range of temperatures likely to be encountered with patients. To enable readings to be easily made the stem of the thermometer is shaped so that in the direction in which the scale is viewed it acts as a magnifying glass.

Liquid-in-metal thermometers

Liquid in metal thermometers (see Fig. 2.24) are more robust than liquid-in-glass thermometers. The thermometer consists of a metal bulb connected to a Bourdon tube. The arrangement is filled with a suitable liquid. When the liquid expands it causes the Bourdon tube to straighten out to some extent and thus can cause a pointer to move across a scale.

Change in temperature
$\longrightarrow$ expansion of liquid
$\longrightarrow$ straightening out of Bourdon tube
$\longrightarrow$ movement of free end of Bourdon tube
$\longrightarrow$ movement of pointer across scale

With mercury the range is −39°C to 650°C, with alcohol −46°C to 150°C. Accuracy is about ± 1% of full-scale reading.

Gas-in-metal thermometer

When the temperature of a gas increases two things can happen. If the gas is allowed to expand freely then it can increase in volume without any increase in pressure. However, if the gas is not allowed to expand, because it is in some container, then no change in volume means the pressure increases. Measurement of the change in volume when the pressure is constant, and measurement of the change in pressure when the volume is constant, can both be used as the basis of thermometers. The most commonly used one for industrial purposes is the change in pressure when the volume is kept constant.

The industrial form of a gas thermometer is usually a thermometer bulb connected to a Bourdon gauge (like the liquid-in-metal thermometer described in Fig. 2.24) and filled with gas, e.g. nitrogen. When the temperature rises the gas pressure increases and this causes the Bourdon tube to straighten out to some extent and can cause a pointer to move across a scale.

Change in temperature
　　⟶ change in pressure of gas
　　　⟶ straightening out of Bourdon tube
　　　　⟶ movement of free end of Bourdon tube
　　　　　⟶ movement of pointer across a scale

The bulb of the thermometer is fairly large, about 50–100 cm^3. The thermometer is robust, has a range of about $-100\,°C$ to $650\,°C$, is direct reading, can be used to give a display at a distance, and has an accuracy of about $\pm\,0.5\%$ of full-scale deflection.

Vapour pressure thermometers

The amount of evaporation from a liquid depends on the temperature; the higher the temperature the greater the amount. Evaporation means that molecules of the liquid are escaping from the liquid into the space above the liquid. These escaped molecules exert a pressure called the *vapour pressure*. This pressure is thus a measure of the temperature of the liquid.

The *vapour pressure thermometer* consists of a thermometer bulb connected to a Bourdon gauge (in a form just like the liquid-in-metal thermometer described in Fig. 2.24) and containing a small amount of liquid. The higher the temperature the greater the amount of liquid that has evaporated and the greater the pressure exerted by its vapour. A higher pressure causes the Bourdon tube to straighten out to some extent and can cause a pointer to move across a scale.

Increase in temperature
　　⟶ increase in evaporation of the liquid
　　　⟶ increase in vapour pressure
　　　　⟶ straightening out of Bourdon tube
　　　　　⟶ displacement of free end of tube
　　　　　　⟶ movement of pointer across scale

With methyl chloride as the liquid the range is about 0–$50\,°C$, with ethyl alcohol 90–$170\,°C$, toluene 150–$250\,°C$. The scale is non-linear and there is an accuracy of about $\pm\,1\%$. The instrument is robust, direct reading and can be used at a distance.

Metal resistance thermometers

The electrical resistance of a coil of metal wire depends on its temperature, the higher the temperature the higher the resistance. The change in resistance depends on:

1 *The initial resistance* The higher the initial resistance of the coil the greater its change in resistance.

2 *The change in temperature* The greater the change in temperature the greater the change in resistance.

3 *The metal used* These factors can be combined to give the equation:

$$\text{change in resistance} = R_0\theta\alpha$$

where R_0 is the initial resistance, θ the change in temperature and α is the temperature coefficient of resistance. This depends on the metal used. The equation indicates that the change in resistance of a coil of wire is directly proportional to the change in temperature. A graph of change in resistance plotted against change in temperature is thus a straight line.

If R_θ is the resistance at temperature θ then the change in resistance is

$$R_\theta - R_0 = R_0\theta\alpha$$

and hence the usual form which is used for the relationship

$$R_\theta = R_0(1 + \alpha\theta)$$

This can be rearranged to give

$$\frac{R_\theta}{R_0} = 1 + \alpha\theta$$

Hence a graph of R_θ/R_0 against θ gives a graph with a slope of the temperature coefficient of resistance. Figure 8.7 shows such graphs for platinum, nickel and copper. Over the temperature ranges indicated the graphs are virtually straight lines and so obeying the above relationship.

Platinum is widely used for *resistance thermometers*. This metal has a closely linear resistance–temperature relationship, gives good repeatability, can be used over a wide range of temperatures (about −200°C to 850°C) and because it is relatively inert can be used in a wide range of environments without deterioration. It is however more expensive than many other metals but the benefits outlined above tend to outweigh the cost factor. The temperature coefficient of resistance α is about 0.0039 /°C.

Nickel and copper are cheaper alternatives but are more prone to interaction with the environment and cannot be used over such a large range of temperature. Nickel has a temperature coefficient of resistance α of about 0.0067 /°C and a range of about −80 to 300°C. Copper has a temperature coefficient of resistance α of 0.0038 /°C and a range of about −200 to 250°C.

Whatever the metal used, the resistance element generally consists of the resistance wire wound over a ceramic coated

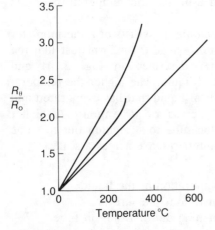

Fig. 8.7 Resistance–temperature graphs for platinum, nickel and copper

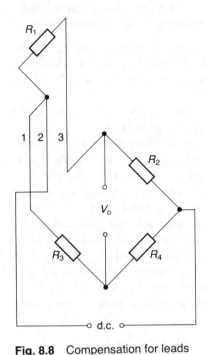

Fig. 8.8 Compensation for leads

tube, it then being also coated with ceramic, and mounted in a protecting tube. The result is a probe for immersion in the medium whose temperature is being measured. The response time is fairly slow, often of the order of a few seconds, because of the poor thermal contact between the coil and the medium whose temperature is being measured.

Resistance thermometers with metal wire coils are generally used with a Wheatstone bridge (see Ch. 3). A problem with just including the resistance element in one arm of the bridge is that the resistance measured will be that of the resistance coil and the leads connecting it to the bridge. If the temperature of the leads changes then there will be a resistance change, regardless of what happens to the temperature of the resistance element. One way of avoiding this is to use three leads for the resistance element and connect them to the bridge in the way shown in Fig. 8.8. Any change in resistance of lead 1 is added to the resistance R_3. Any change in resistance of lead 2 has no effect on the balance conditions of the bridge since that lead is in the d.c. supply connection to the bridge. Any change in resistance of lead 3 is added to the resistance of the resistance element R_1. Thus changes in the resistances of the leads affects equally two arms of the bridge and will cancel out if R_1 and R_3 are about the same resistance.

Example 2

Under what circumstances would a nickel resistance thermometer be preferred to a platinum resistance thermometer?

Answer

Nickel is to be preferred where a larger change in resistance per degree change in temperature is required. It is also cheaper.

Semiconductor resistance thermometers

Resistance thermometers using thermistors (see Ch. 2) give much larger resistance changes per degree than metal wire elements. However the variation of resistance with temperature is not linear. Their small size means a small thermal capacity and hence a rapid response to temperature changes. The temperature range over which they can be used will depend on the thermistor concerned, ranges within about $-250\,°C$ to $650\,°C$ are possible. There is a tendency for the calibration to change with time. A Wheatstone bridge can be used for the resistance measurement.

Thermocouples

A thermocouple consists of two different metal wires in contact, as in Fig. 8.9(*a*). An e.m.f. is produced between the two junctions when there is a temperature difference between the two junctions (see Ch. 2 for more discussion of this). There are a number of laws relating to the thermocouple:

- The thermoelectric e.m.f. depends on the temperatures of the two junctions. Usually one junction is kept at 0°C and so then the e.m.f. depends on the temperature in °C of the other junction.
- The temperatures of the wires between the two junctions has no effect on the e.m.f. (Fig. 8.9(a)).
- A thermocouple circuit can have other metals in the circuit and they will have no effect on the thermoelectric e.m.f. provided all their junctions are at the same temperature (Fig. 8.9(b)). Thus, for example, a voltage-measuring instrument can be introduced into the circuit.
- The e.m.f. given by a thermocouple with junctions at θ_1 and θ_2 is the same as would be produced by adding together the e.m.f.s of two thermocouples, of the same metals, with junctions of one at temperatures θ_1 and θ_3 and the other at temperatures θ_3 and θ_2 (Fig. 8.9(c)). This relationship is known as the *law of intermediate temperatures*.

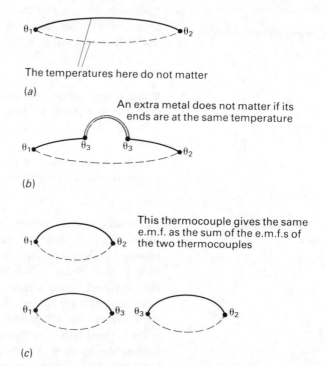

The temperatures here do not matter

(a)

An extra metal does not matter if its ends are at the same temperature

(b)

This thermocouple gives the same e.m.f. as the sum of the e.m.f.s of the two thermocouples

Fig. 8.9 Basic thermocouple laws   (c)

Commonly used thermocouples are shown in Table 8.3 with the temperature ranges over which they are generally used and typical sensitivities. These commonly used thermocouples are given reference letters. For example the iron–constantan thermocouple is called a type J thermocouple.

Table 8.3 Thermocouples

Type	Materials	Range °C	Sensitivity μV/°C
E	chromel–constantan	0 to 980	63
J	iron–constantan	−180 to 760	53
K	chromel–alumel	−180 to 1260	41
R	platinum–platinum/rhodium 13%	0 to 1750	6
T	copper–constantan	−180 to 370	43

The base metal thermocouples, E, J, K and T, are relatively cheap but deteriorate with age. They have accuracies which typically are about ± 1 to 3%. The noble-metal thermocouples, e.g. type R, are more expensive but more stable with longer life. They have accuracies of the order of ± 1% or less.

Standard tables are available which give the e.m.f.s of commonly used thermocouples when one junction is at 0°C. Table 8.4 is an extract from such a table.

Table 8.4 Thermocouple tables

Temperature in °C	Thermoelectric e.m.f. in mV				
	Type E	Type J	Type K	Type R	Type T
0	0.000	0.000	0.000	0.000	0.000
10	0.591	0.507	0.397	0.054	0.391
20	1.192	1.019	0.798	0.111	0.789
30	1.801	1.536	1.203	0.170	1.196
40	2.419	2.058	1.611	0.231	1.611
50	3.047	2.585	2.022	0.295	2.035
60	3.683	3.115	2.436	0.361	2.467
70	4.329	3.649	2.850	0.429	2.908
80	4.983	4.186	3.266	0.499	3.357
90	5.646	4.725	3.681	0.571	3.813
100	6.317	5.628	4.095	0.644	4.277

It is usual to keep one junction of the thermocouple at 0°C. This can be done by immersing it in a mixture of ice and water. An alternative to this is to include in series with the thermocouple a circuit which gives a potential difference which just compensates for the junction being at some other temperature than 0°C.

If one of the thermocouple junctions is not at 0°C then the values given by the tables need modification. This can be done using the law of intermediate temperatures (see earlier). According to this:

e.m.f. for thermocouple with junctions at θ_1 and θ_2
= e.m.f. of thermocouple with junctions at θ_1 and θ_3
+ e.m.f. of thermocouple with junctions at θ_3 and θ_2

To illustrate this consider an iron–constantan thermocouple (type J). The e.m.f. of such a thermocouple with one junction at 100°C and the other at 0°C is the e.m.f. of the thermocouple with one junction at 100°C and the other at 20°C plus that of a thermocouple with junctions at 20°C and 0°C. Hence

> e.m.f. for thermocouple with junctions at 100°C and 20°C
> = e.m.f. of thermocouple with junctions at 100°C and 0°C
> − e.m.f. of thermocouple with junctions at 20°C and 0°C.

Hence, using the data given in Table 8.4,

> e.m.f. of thermocouple with junctions at 100°C and 20°C
> = 5.628 − 1.019 = 4.609 mV

In some instances a group of thermocouples are connected in series so that there are perhaps ten or more hot junctions sensing the temperature. The e.m.f.s produced by each are added together. Such an arrangement is known as a *thermopile* (Fig. 8.10).

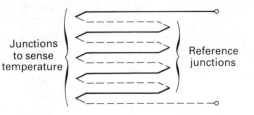

Fig. 8.10 A thermopile

Example 3

Using Table 8.4 determine the e.m.f. that would be produced by a chromel–constantan thermocouple with one junction at 0°C and the other at 70°C.

Answer

Using the table, the e.m.f. is 4.329 mV.

Example 4

A copper–constantan thermocouple is found to produce an e.m.f. of 1.820 mV when one of its junctions is at 0°C. What is the temperature of the other junction? Use the data given in Table 8.4.

Answer

For the copper–constantan thermocouple an e.m.f. of 1.611 mV is produced at 40°C and 2.035 mV at 50°C. The temperature therefore must be between 40°C and 50°C. Over this temperature range a one-degree change produces an e.m.f. of

$$\text{e.m.f. per degree} = \frac{2.035 - 1.611}{50 - 40} = 0.0424 \text{ mV}$$

Therefore the number of degrees above 40°C that the temperature is will be

$$\frac{1.820 - 1.611}{0.0424} = 4.929\,°C$$

Hence the temperature is 44.929°C.

Radiation pyrometers

We can feel the radiation emitted by the red-hot element of an electric fire. The hotter the element the more radiation it emits. The radiation emitted per second by an object depends on the temperature of the object. With the *radiation pyrometer* the radiation from the object is focused onto a radiation detector (Fig. 8.11). The radiation detector might be a resistance thermometer, a thermistor, or a thermopile (this is a group of thermocouples in series). The output from such a detector is a measure of how much the radiation has raised its temperature. The output is thus a measure of the temperature of the object emitting the radiation.

An object at temperature
⟶ radiation, the amount being related to the temperature
⟶ raises the temperature of the detector
⟶ detector gives an output

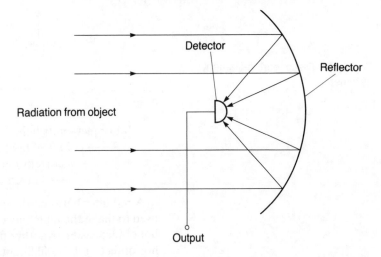

Fig. 8.11 Radiation pyrometer

Radiation pyrometers have the great advantage that they do not have to be in contact with the object whose temperature is being measured. They can thus be used for objects which are too hot for contact, too corrosive, or moving. They can also be used for surfaces since they do not have to be inserted into the hot object. They can be used for temperatures from about room temperature to 1800°C.

Disappearing-filament pyrometer

We frequently talk of hot objects being red hot or if hotter still white hot. The colour of the light emitted by a hot object depends on its temperature. The *disappearing-filament pyrometer* (Fig. 8.12) uses colour as a measure of temperature. The instrument has the radiation focused onto a filament so that the radiation and the filament can both be viewed in focus through an eyepiece. The filament is heated by an electrical current being passed through it. The current through the filament is raised until the filament and the hot object seem to be the same colour, i.e. the filament disappears into the background of the hot object. When this occurs the filament current is a measure of the temperature.

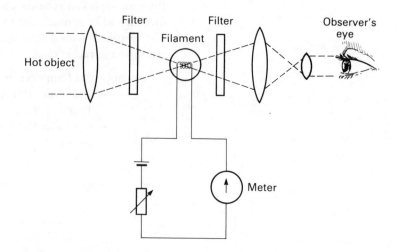

Fig. 8.12 Disappearing-filament pyrometer

Hot object emits light
    ⟶ colour of light matched with colour of hot filament
      ⟶ current determines filament temperature
      ⟶ current registered on a meter

A red filter between the eyepiece and the filament is generally used to make the matching of the colours of the filament and the hot object easier. Another filter may be introduced between the hot object and the filament. This has the effect of making the object seem less hot in comparison with the filament and so extends the range of the instrument. The disappearing-filament pyrometer can be used from about 500°C upwards and has an accuracy of about ± 0.5% of the reading.

Problems

1　Explain the basis on which the International Practical Scale of temperature is defined.

2　Why are melting points and boiling points chosen for the fixed points of the International Practical Temperature Scale?

3　A mercury-in-glass thermometer is said to be calibrated for partial immersion. What does this mean?

4　A chromel–constantan thermocouple has one junction at $0\,°C$ and the other in a hot liquid. Estimate the temperature of the liquid if the thermocouple gives an e.m.f. of 3.320 mV. Use the table given in this chapter.

5　A nickel resistance element has a resistance of $100.0\ \Omega$ at $0\,°C$. What is the temperature when its resistance becomes $109.5\ \Omega$? The temperature coefficient of resistance of nickel is $0.0067\,/°C$.

6　What advantages does a bimetallic thermometer have in comparison with a mercury-in-glass thermometer?

7　Why do bimetallic thermometers frequently use a spiral or a helix strip rather than just a straight bimetallic strip?

8　Explain how the effect of temperature on the resistance of the leads to a resistance thermometer can be eliminated.

9　Suggest measurement systems that could be used to determine temperatures in the following situations:

(a) A liquid at about $60\,°C$ and the results to be presented some short distance away.

(b) The air temperature in the room, about $20\,°C$.

(c) The surface temperature of a sheet of steel, about $400\,°C$.

(d) The temperature of a small sample of material, about $150\,°C$.

10　Explain the principles of operation of the following temperature measurement systems:

(a) Bimetallic thermometer

(b) Vapour pressure thermometer

(c) Metal resistance thermometer

(d) Disappearing-filament pyrometer.

11　Resistance thermometers are widely used in the process industry for monitoring the temperature of liquids. What are the merits of resistance thermometers for such situations?

12　(a) Design a temperature-measurement system for a liquid using a thermistor. The temperature range is 15–70 °C and an accuracy of about $\pm\,0.5\,°C$ is required. You should consult manufacturer's data sheets in order to select a suitable thermistor.

(b) Set up your system, test and calibrate it.

9 Maintenance

Failure rate

Failure means that a component or system is no longer able to do the job it was intended for. Failure may be partial in that the characteristics of the item deviate from the required specification but not so much as to cause complete inability to do the intended job. Thus, for example, a resistor's resistance may change as a result of the conditions under which it is being used by more than is specified for a circuit and so change the calibration of an instrument. Complete failure is when the deviation is so great as to mean that the item can no longer carry out the required job. Failure may arise as a result of misuse, environmental conditions or an inherent weakness in the item itself. Thus, for example, an electronic component may fail because the applied voltage was too high, it was subject to mechanical vibrations or shock, or mechanical stresses were produced in making the connections to the component.

The *failure rate* of a component can be defined as the percentage of failures that can be expected to occur over a specified period of time or number of uses.

$$\text{Failure rate} = \frac{\text{number failing}}{\text{number observed}} \times 100\%$$

The failure rate can be found for a component or a system by taking a large number of them and finding the number of failures when they are repeatedly used or operated over a period of time. For example, tests on an electrical resistor might show that of 5000 tested by being in continuous use 12 failed in 2000 hours. The failure rate in that period of time is thus

$$\text{failure rate} = \frac{12}{5000} \times 100\% = 0.24\%$$

Figure 9.1 shows how the failure rate of components typically varies with time. Because of its shape the graph is frequently referred to as the *bath-tub graph*. The graph shows three distinct periods. During the infancy or *burn-in* period there is a relatively large failure rate caused by manufacturing faults or the effects of installing the components. After that period the components settle down to a relatively stable middle age. Failures during this period are fairly random. The third period is old age when the failure rate increases as a result of components wearing out.

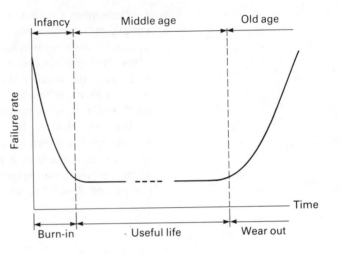

Fig. 9.1 The bath-tub graph

The purpose of maintenance

The purpose of maintenance is to:

- keep systems in a condition that ensures they operate effectively and efficiently
- keep the costs low of 'lost time' due to breakages or mis-functioning systems

The term *preventive maintenance* is used to describe maintenance which is used to delay or prevent breakdown. It involves inspection to determine which components need replacing or repair before they reach the breakdown point. Preventive maintenance is a policy of replacing components that are nearing the end of their life before they actually wear out and fail. It is a policy of anticipating failure. It also involves servicing, i.e., routine cleaning, lubrication, adjustment, etc., to reduce wear and hence prevent breakdown. It is necessary with such maintenance to keep maintenance records, indicating when inspection, servicing, repairs and replacements were carried out. Such records are necessary to ensure that servicing has been carried out at the correct times and to enable components to be replaced just before they can be expected to wear out.

The term *corrective maintenance* or *breakdown maintenance* is used to describe a maintenance policy which is to replace or deal with failure only when it occurs. It involves the detection, location and repair or replacement of faulty items as they occur. It is a policy of waiting for failure before taking any action.

Maintenance costs

The total cost of maintaining a system can be considered to be made up of the costs of:

1 *Preventive maintenance* The greater the amount of preventive maintenance the greater the cost.

2 *Breakdown maintenance* The greater the amount of preventive maintenance the smaller the number of breakdowns and hence the smaller the cost of breakdown maintenance.

3 *Lost time* When an item breaks down it can no longer be used while it is repaired or replaced. This might mean a production process has to be stopped. Such a loss in production has a cost. There are thus costs associated with the time lost as a result of a breakdown. The greater the amount of preventive maintenance the smaller the number of breakdowns and hence the smaller the lost time costs.

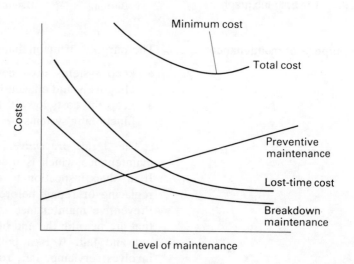

Fig. 9.2 Maintenance costs

Figure 9.2 shows typical forms graphs of the above costs might take. The total maintenance cost is the sum of the costs of the three elements. The total cost graph shows a minimum at a particular level of maintenance. This is obviously the level of maintenance to adopt to ensure minimum costs.

Life-cycle costs

Life-cycle costs are the total costs of an item throughout its entire life. This includes its initial cost, its maintenance cost and any other costs involved in keeping it going. In trying to keep the life cycle costs down to a minimum consideration has to be given to:

- how long the item will last;
- how reliable it is, i.e., how often it is likely to break down;
- how much maintenance it requires;
- how long it will take to carry out maintenance;
- how easy it is to get spare parts.

In designing the item consideration has to be given to not only the need for the item to carry out the required job but also for reliability and ease of maintenance. Installing and commissioning procedures, e.g., checking the correct functioning of items and calibration, need planning to ensure that items are correctly used and so have the maximum life possible. Maintenance needs planning to ensure that it is at minimum cost. The term *terotechnology* is used for all the activities involved in trying to keep life costs down to a minimum.

Maintenance manuals

The maintenance manual for a system is likely to include the following features:

- A description of the system and its use
- A performance specification
  This would include such details as the ranges of an instrument, its accuracy and the conditions under which it is expected to be used.
- Calibration details
  Many measurement systems not only require initially calibrating, either by the manufacturer or during commissioning, but also frequent calibration checks afterwards to check that the system is still correctly calibrated. Calibration records showing how the calibration has changed with time are useful ways of indicating when components are wearing out and preventive maintenance is required.
- Block diagrams showing the various elements included in the system and how they are linked.
- A mechanical layout
  This could be in the form of diagrams or photographs showing the location of the various elements.
- Circuit diagrams
  These would give details of the electrical circuits of each of the constituent electrical elements of the system and details of each of the electrical components used.

- Maintenance procedures
  This could include preventive maintenance procedures, e.g., oiling procedures, replacement of components after certain times. It also could include a fault location guide and test procedures.
- Spare parts list

Examples of maintenance procedures

The following are some of the maintenance procedures that are likely to be part of the servicing requirements for measurement systems.

Pressure measurement

Instruments with mechanical linkages, e.g. Bourdon gauges, can undergo wear and misalignments. Inspection can reveal such problems. Dirt can affect the operation of linkages. Cleaning and suitable lubrication can ensure that linkages operate smoothly.

The piping used for pressure systems can become clogged with dirt; also leaks can develop. Inspection and regular cleaning of piping can be a necessary service requirement.

Manometers and barometers employing mercury can have their measurements significantly affected by dirt contaminating the mercury. A service requirement may thus be periodic inspection of the mercury, and possibly replacement by clean mercury.

Level measurement

Sight glasses may become dirty and make estimation of level difficult. A service requirement is thus likely to be the periodic closure of the arm containing the sight glass so that it can be removed and cleaned.

The service requirements for float-operated instruments can be inspection for corrosion of the float and any other parts of the instrument in contact with the liquid, lubrication of moving parts such as pulley wheels, inspection to check that tapes or wires attached to floats are not wearing, broken or twisted.

Pressure measurement systems used for level measurements need to be inspected as indicated above for pressure measurement. There is a particular need to inspect for leaks. With the bubbler method the main need is probably for cleaning of the tube since it can easily become blocked.

Flow measurement

Inspection for leaks in piping, wear at orifices and silting, i.e., deposits of solids behind such items as an orifice plate, are likely to be routine service requirements. Pressure measurement systems associated with flow measurement will require the type of servicing suggested above.

Temperature measurement

Instruments employing Bourdon gauges will require the type

of servicing indicated for pressure measurement systems. Another requirement may be for calibration checks. This can be done by comparing the responses with those of a standard thermometer.

Problems

1 Define the term *failure rate* and explain how the rate is likely to change with time.
2 Explain the difference between preventive and breakdown maintenance.
3 Explain the purposes of inspection and servicing of systems.
4 Explain the significance of the following in relation to costs:
   (a) preventive maintenance,
   (b) designing for maintenance,
   (c) breakdowns.
5 Use a maintenance manual to check through a measurement system.
   (a) Identify the various constituent elements and their functions.
   (b) Check all calibration points.
   (c) Carry out any servicing and consider whether any other preventive maintenance is required.
6 Design a measurement system for monitoring the temperature in a room/oven/vat of liquid and producing a continuous record. Also design the maintenance manual for the system. In particular, indicate how the system is to be initially calibrated and the calibration later checked.

Answers to problems

Chapter 1

3. 35.4 to 36.2°C
4. 2.8 to 3.6 V
5. (a) 0 to 20 km/h, (b) 20 to 160 km/h
6. (a) +30 kPa, (b) ± 0.6 f.s.d.
7. ± 2.5%
8. About ± 4.7°C
9. 855 ± 10°C

Chapter 2

1. (a) temperature–resistance, (b) temperature–e.m.f., (c) light-intensity–resistance change, (d) strain–resistance change, (e) pressure–straightening out, (f) temperature–volume change
2. For example, (a) LVDT, (b) spring, (c) load cell with strain gauges, (d) resistance thermometer, (e) thermocouple, (f) liquid-in-steel thermometer

Chapter 3

2. (a) Resistance $\rightarrow$ potential difference, (b) e.m.f. $\rightarrow$ rotation, (c) pressure $\rightarrow$ displacement
3. (a) 0.020 V, (b) change to 101.0 Ω
4. 0.339 V
5. (a) small displacement $\rightarrow$ larger displacement, (b) large signal $\rightarrow$ smaller signal, (c) small voltage $\rightarrow$ larger voltage
7. $R_2/R_1 = -30$
8. 0.047 V
9. (a) lever, (b) gear train, (c) amplifier, (d) Wheatstone bridge

Chapter 4

6. 5 mm
7. 25 mm

Chapter 5
2. 16 kPa
6. 100.7 kPa
7. +0.25 mm
8. (a) +0.3 mm, (b) 102.4 kPa

Chapter 6
4. (a) 19.6 N or 2.0 kg, (b) 39.2 N
5. 1.2 kPa
6. (a) conductive probes, (b) ultrasonic or radiation, (c) float
7. Range of levels measured
8. A diaphragm pressure cell

Chapter 7
4. Venturi tube
5. (a) 0.020 m^3/s, (b) 24 kg/s
6. 9.1×10^{-6} m^3/s

Chapter 8
4. 55.09 °C
5. 14.2 °C
9. E.g. (a) liquid or gas-in-metal thermometer, (b) mercury-in-glass thermometer, (c) a radiation pyrometer, (d) thermocouple